Language and Neurology

Series Editor
Patrice Bellot

Language and Neurology

Alzheimer's Disease

Christophe Cusimano

iSTE

WILEY

First published 2021 in Great Britain and the United States by ISTE Ltd and John Wiley & Sons, Inc.

ISTE Ltd
27-37 St George's Road
London SW19 4EU
UK

www.iste.co.uk

John Wiley & Sons, Inc.
111 River Street
Hoboken, NJ 07030
USA

www.wiley.com

Library of Congress Control Number: 2020946491

British Library Cataloguing-in-Publication Data
A CIP record for this book is available from the British Library
ISBN 978-1-78630-662-3

Contents

Acknowledgments

First and foremost, I would like to thank my family for their patience and forbearance, subject as they are to the vagaries of an academic's schedule. To my wife, who has learned to live with my own particular brand of organized chaos, and to my three wonderful children who have given up asking what exactly I do at work.

Warmest thanks to François Rastier for his valuable advice, both in relation to this book and elsewhere, and to Patrice Bellot, director of the Cognitive Science collection at ISTE, for his encouragement and support over the course of this project.

Thanks go to all of the patients who allowed me to sit in on their consultations, and also – especially – to all of the neuropsychologists and speech therapists, too numerous to name, who kindly allowed me to observe their work. This book would not have been possible without their willing participation and valuable insights.

Finally, I wish to thank my fantastic students at Masaryk University in Brno, Czech Republic, who never cease to inspire and motivate me to give my best every semester; my colleagues in the Romanistics department, for their unfailing kindness; and all of my other colleagues and friends at Masaryk.

Introduction

Application of Linguistics to Cognitive Tests and Speech Patterns in Alzheimer's Patients

I seemed to be upon the verge of comprehension without power to comprehend – men, at times, find themselves upon the brink of remembrance without being able, in the end, to remember.

Edgar Allan Poe
The Murders in the Rue Morgue, 1841

Despite playing a major role in laying the theoretical foundations for many of the humanities and social sciences, interest in linguistics has waned over the last 20 years. The decline of structuralism, coupled with the rise of both cognitive sciences and communications theory, means that the subject is increasingly limited to strictly academic circles; even in this restricted context, it can be hard to convince funding bodies of the utility – that all-important, overriding criterion – of linguistics projects. It would be wise to note the significant role played by linguistics in the fields of natural language processing (e.g. in search engine development, automatic translation, etc.) and in foreign language teaching, where it plays a supporting, but no less essential, role.

In this book, we have chosen to focus on a field of application in which linguistics plays a supporting role, providing theoretical insights into a domain of expertise. The application of linguistics to the field of medicine is relatively novel, and not without its risks; success is not guaranteed. Jakobson highlighted this issue in his 1963 conference presentation on aphasia (1969, p. 133): "I make no claim to expertise in pure linguistics, psychology or medicine; as such, my comments will be limited to linguistic observations of linguistic facts, and no more". Similarly, in this book, we have chosen to focus on two aspects:

1) The detection of Alzheimer-type dementia (and, consequently, of certain related conditions) using cognitive testing. We will not go into any detail concerning other methods of investigation, whether of the *phenomenological* type (perception through oral questioning or questionnaires) or of the strictly *experimental* type. Instead of the head-on approach used by neuropsychologists, speech therapists and geriatric specialists, we have chosen to view the problem from a different angle, focusing on providing support for diagnosis. We shall concentrate on one simple question, asking whether or not language-based cognitive tests are well-designed in linguistic terms, and whether they provide sufficient information. We ask whether these tests are supported by findings from the field of linguistics, particularly with regard to textual semantics. Evidently, even if this is not the case, it is not our intention to pass judgment on the practices of healthcare professionals.

2) An investigation of the linguistic characteristics of speech in patients suffering from Alzheimer's disease. This more "conventional" study aims to provide a thorough inventory of traces of amnesic troubles in patient speech, and of the strategies used by sufferers to work around these problems.

This book follows a relatively predictable plan, beginning with a review of the limited number of existing studies which consider various pathologies from a linguistic perspective. Next, we shall give an overview of the linguistic symptoms of Alzheimer's disease, touching briefly on other related conditions. Chapter 3, which forms the basis for the rest of our work, consists of a detailed examination of diagnostic tests, beginning with general tests before moving onto those which focus more specifically on speech in the context of memory-related pathologies. The majority of our critical remarks, along with suggestions for improvements, are found in this section. We highlight a number of promising new pathways, while giving further consideration to certain ideas which have only briefly been touched upon in the literature and which, in our view, merit further investigation.

For instance, the textual dimension of information encoding produces a large volume of recall cues, but is systematically excluded from test procedures in favor of a less "ecological", more symptomatic and more "evocative" approach. Similarly, cultural differences are generally ignored and left open to interpretation by the practitioner. Our intention is to carry out a more in-depth investigation of these possible pathways, although we make no pretense to verify these hypotheses in practice or produce new tests as a function of our findings. Chapter 4 of this book concerns recent developments in the domain of patient speech, touching on the notions of corpus linguistics and on the studies of interviews carried out with Alzheimer's patients in a clinical context. This chapter includes several extracts from interviews which we carried out: in these transcripts, we attempt to identify characteristic features of patient speech, relating to syntax, semantics or other areas. Due to our own professional background, we focus on this latter consideration, particularly in terms of confabulation analysis.

This book contains the critical reflections of a linguist concerning a field of application which, at first glance, lies far outside his area of expertise. In presenting these observations, we aim to stimulate reflection and debate on the subject, with the aim of improving the lot of patients. Our experience of working alongside neuropsychologists and speech therapists has led us to believe that such progress is eminently possible. Evidently, the ultimate goal of our work is to contribute to the improvement of diagnostic tests, using findings from linguistics to fine-tune practical approaches. Furthermore, we hope that our speech analyses will help caregivers to better support and better understand patients, improving communication. This work may be of interest to non-specialist readers who have encountered cognitive testing procedures in the past (e.g. friends and relatives of dementia patients) or who simply wish to know more about specific characteristics of the speech of Alzheimer's sufferers; to this end, we have chosen to write in a very direct style with as little jargon as possible. As we shall see, this book makes no claim to exhaustivity, and our analyses may lay the foundation for future work, finally responding to the decades-old call issued by Hatfield (1972): "Looking for help from linguistics".

Conversation transcription conventions

Evidently, our purpose in writing this book is not simply to provide a collection of patient speech extracts for researchers working on interaction and conversation analysis. As such, we have chosen to prioritize readability for non-specialists, using minimal transcription conventions based on Jefferson's grid (2004), as used by Asp and De Villiers (2010, p. vi).

C	Caregiver
P	Patient
S	Sufferer
NP	Neuropsychologist
ST	Speech therapist
(1)	Line 1
(10)	Line 10
( )	Audible but unclear
(( ))	Clarification or description, e.g. ((laughs))
#	Short pause in a speech turn
##	Longer pause in a speech turn
// //	Breath group, i.e. group of words pronounced by the speaker without stopping for breath, ending with a pause
xxx	Inaudible
^	Short silence between two speech turns
^^	Longer silence between two speech turns

Table I.1. *Conversation transcription conventions*

1

Linguistics, Language Pathologies and Alzheimer's Disease: A Brief History

As we noted in the introduction, very few linguists have studied language pathologies in a medical context; the development of specialized linguistic theories with practical applications in this area has seemingly been perceived as overly challenging. This does not mean that theories developed in unrelated contexts – for example, in our own field of interpretative semantics or speech analysis – may not be used to cast light on these issues; however, it reinforces the general perception of a persistent lack of research in this area.

It is thus the history of the domain itself, and not simply our summary of this history, which is "brief". The lack of interest in this area on the part of linguists has left the field wide open for a series of postulates, mostly derived from the field of neuroscience, to take root; we shall discuss these postulates in greater detail later.

For the moment, we shall focus our discussion on the *theory of mediation*, the only linguistic theory focusing on language pathologies; we shall then turn our attention to other theories which have been used in attempting to describe or explain these pathologies.

1.1. Gagnepain's theory of mediation: an approach to pathological speech on its own terms

Few linguistic theories have addressed the issue of language disorders. One of the most incendiary and original theories in this select group is Jean Gagnepain's theory of mediation. This theory is, inexplicably, not well known among linguists; however, its wide applicability has inspired many researchers and supporters from across the field of social sciences. The theory has its roots in a meeting between Gagnepain and Sabouraud, a neuropsychiatrist, following the former's installation as professor of "pure" linguistics at Rennes in 1956. At this early date, there was nothing to suggest that Gagnepain – whose thesis concerned Celtic syntactics – would direct his attentions to pathologies. This change in direction was to his advantage: the author is not particularly well known for his earlier work[1], but for his proposed "deconstruction" of language, which reverses the usual logic whereby we define a norm, then describe deviations from this norm. On the contrary, Gagnepain and Sabouraud considered language pathologies as a type of reagent, revealing the true, underlying mechanisms of language; each pathology highlights, in some way, processes which are not visible when everything is functioning "correctly" (Gagnepain 2006, p. 54)[2]:

> "We consider the mental patient as a teacher of linguistics: he or she teaches us how language works. Insofar as its operation is not haphazard, we attempt, from the patient's behavior, to induce the grammatical rules which govern the construction of their messages"[3].

1 Gagnepain himself, later in his career (February–March 1993) (Gagnepain 2016, p. 18) clearly recognized the true value of his background in linguistics: "I mention this approach to linguistics now to avoid having to speak of it later; I believe that if one wishes to study the humanities and social sciences, one must rid oneself of all these old clichés which either establish divisions between man and "natural" objects, or attempt to 'naturalize' man".

2 Author's note: many of the quotations of this book have been translated from the original French. For reasons of clarity, the original text will not be given; interested readers may wish to consult the bibliography.

3 Note that this idea was not entirely original; the role of exceptions in illustrating the operation of a system is well known. As regards the brain, the great science-fiction writer Asimov, in a short story (*Stranger in Paradise*, 1974), wrote, of an autistic character: "Randall, subjected for increasing lengths of time to artificial stimuli, yielded up the inner workings of his brain and gave clues thereby to the inner workings of all brains, those that were called normal as well as those like his own".

Before going into detail concerning this theory and its main contributions, it is interesting to note that the dominance of structuralism in recent decades has resulted in the marginalization of certain "fringe" theories in spite of their clear structuralist slant, such as Tesnière's structural syntax. The same can be said of the theory of mediation, situated by its authors along these lines (Tesnière 1963, p. 86), as we clearly see from the conclusion to their remarkable study of aphasia:

> "Using structural linguistics, we have attempted to define the principal illustrations provided by the study of aphasic patients".

The *theory of mediation* (TDM, from the French *théorie de la médiation*) posits that human beings exercise a form of "mediation" in their relations with the world. Gagnepain defines this as "the intrusion, into the immediate of an element which cannot be immediately understood, of a mediate, whether in the structure, in praxis, or on an implicit level" (Gagnepain 2016, p. 31). In this quotation, we see that the author refuses to assimilate reason to language alone; we also detect the "essence" of the TDM itself, in the division of human capacities into a series of *planes*, or levels. The TDM is systematically referred to as a theory of *deconstruction*, by Gagnepain himself and by subsequent commentators. This deconstruction is rendered necessary by the number and variety of disorders observed in patients, and operates on four different planes, not just the capacity of speech; Gagnepain sweepingly asserted that all previous authors – including Saussure – had limited their reflection to this more simplistic approach:

> "We have made a clear break with those who have gone before us in refusing to reduce rationality to a single mode. While most thinkers have restricted reason to its verbal form alone, following the Greek tradition of the *logos*, we have established four rational modes".

According to Gagnepain, these four planes are not *a priori* objects, but rather *products*, based on observations. Working closely with Sabouraud, Gagnepain gained access to a large group of aphasic patients (the subjects of the authors' earliest analyses), along with dyslexic, dysgraphic, dyspraxic and even schizophrenic patients, straddling the line between neurology and psychiatry with ease. This approach to pathologies as a whole is the reason TDM is also referred to as a "clinical anthropology".

The difficulties and inabilities of each type of patient highlight different capacities. Mjachký (2016, p. 29), author of a thesis on determinants which was based, in part, on the theory of mediation, calls attention to both the independence of planes (enabling understanding of phenomena) and the unity resulting from their presence, to a greater or lesser extent, in all subjects:

> "These modalities, which Gagnepain refers to as 'planes', are 'language', the aforementioned faculty of representation; the 'tool', which is the faculty of fabrication; the 'person', which is the faculty of the social being; and the 'norm', which is the faculty of ethical analysis. The value of this model lies, amongst other aspects, in the fact that each of these autonomous rationales is viewed analogically, i.e. that man, whilst a pluridimensional being – deconstructed, in a certain way – remains one, precisely in light of the identity of construction of each dimension (plane)".

The four planes identified by Gagnepain and Sabouraud, in their own terms, are: language (plane I, giving rise to the science in question here, *glossology*); tools (plane II, the subject of *ergology*); the person (plane III: *sociology*, and to some extent *politics* and even *legislation*) and norms (plane IV, relating to *ethics*). In this case, Gagnepain showed a rare degree of humility in noting that, as these four planes were identified through observations, his list was not necessarily exhaustive.

Just as the TDM is broken down into four planes, glossology itself can be broken down, in the context of pathology studies, into four levels. These do not correspond in any way to the conventional division of domains within the language sciences; Gagnepain's aim was for these four levels to correspond, analogically, to the four higher levels. In this sense, the term "sub-levels" is not entirely appropriate, as we shall see. This difficulty is clearly explained by Jean-Quentel and Beaud (2006, p. 39):

> "We have thus 'deconstructed' language into four registers which correspond to the four planes of the theory of mediation: grammaticality, writing, word usage and speech. Only one of these four registers, grammaticality, specifically concerns

language. The other three registers result from the action of external processes of a different order on language. Writing is, at the end of the day, a technical matter; word usage relates to a social dimension, and speech involves an ethical process".

The analogy between the divisions at the higher level and those identified within the language register is remarkably original, practical and, according to the author, applicable to a range of pathologies. This double deconstruction is summarized in the diagram below (according to our interpretation: neither Gagnepain nor any of his followers have used this representation).

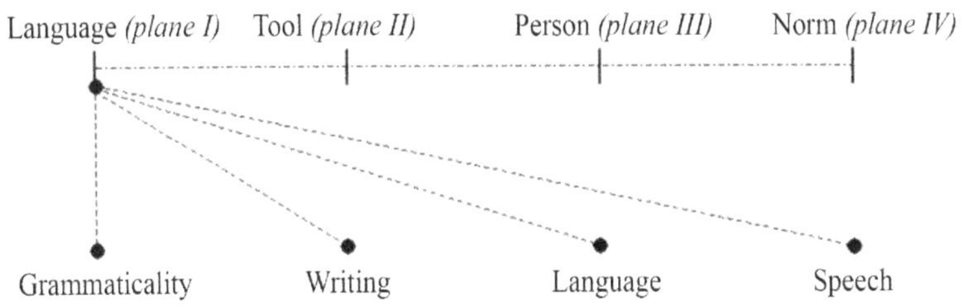

Figure 1.1. *The four planes of the theory of mediation and the four registers of language*

Quentel and Beaud's illustration of language deconstruction does not specifically touch on pathologies, but is centered on a convincing discussion of language acquisition in children. In a succinct discussion, the authors present at least one disorder associated with each dimension. Quentel and Beaud present dysphasia as a disorder relating to grammaticality, and dyslexia as a disorder relating to reading. Disorders connected with word usage are harder to identify clearly, corresponding to infantile psychoses which act on the child's capacity to interact with others but do not affect grammaticality. Finally, communication disorders, or the lack of desire to communicate, relate to the speech register.

Gagnepain and Sabouraud began by testing their hypotheses in the context of aphasia, notably confirming, in practice, Saussure's theory of the biaxiality between paradigm and syntagm. Patients with Wernicke-type aphasia were shown to be unable to complete identification, recognition or choice tasks on both the signifier and signified levels. Patients suffering from Broca's aphasia were unable to segment speech chains in any appropriate way (neither phonemes nor lexical units). Once again, this disorder concerns both the signifier and signified levels. However, Wernicke-type aphasia clearly involves a disorder on the paradigmatic operation level (or the "taxonomic" level, according to Gagnepain 2016, p. 61): "Patients with Wernicke's aphasia have lost the capacity to choose, but retain the ability to segment text; everything in their lexicon is confused". Broca-type aphasia patients cannot carry out syntagm segmentation (Gagnepain's "generative" level, 2016, p. 62): "Patients with Broca's aphasia, who are unable to combine and who cannot add one and one, tend to make more precise choices". These examples both concern language production, but these disorders may also be observed in language reception.

As we have just seen, aphasia sufferers may be grouped according to their relation to biaxiality; they may also be categorized according to the *signifier/signified* duality: "one finding of aphasiology concerns the 'breaking up' of signs: some patients possess the capacity for phonological analysis alone, without semiological analysis, or vice-versa" (Gagnepain 2016, pp. 55–56). The author refers to *phonological* Wernicke's or Broca's aphasia when the capacity for phonological analysis is deficient, and to *semiological* Wernicke's or Broca's aphasia when the capacity for semiological analysis is lacking. In this way, Gagnepain defines four different types. We shall not go into any detail concerning Gagnepain's studies of aphasia here; interested readers may wish to consult the works cited in the bibliography.

Gagnepain only briefly touches on Alzheimer's disease, perhaps due to a lack of personal interest, or to the period in which he was working (1970s–1990s), when the contours of the disease were only just becoming clear. He mentioned the disease only in the context of a discussion of the way in which society regards the elderly. In his sixth "lesson", *Devenir ce que nous sommes* (Becoming What We Are), the author considers elder care homes, offering up a reflection in which he questions the practice of placing seniors in these establishments, noting that the idea of the "senile senior" is a very recent development: "in the past, the old person was still the master of the house; elders were respected as such, even if they had lost their faculties" (Gagnepain 2016, p. 207). Further on, Gagnepain noted that "in the majority of civilizations, before our society 'medicalized' old age, the elder was precisely the one who could not die, was ageless". Our Western societies have an increasing tendency to marginalize the elderly. Coming from a different angle – that of pure psychiatry, or, to use the author's own term, "psychogeriatrics" – Ploton (2004, p. 81) made a very similar remark:

> "In seeking a definition of old age, based on observations, we note, at a more or less advanced stage, the presence of a 'tipping point' where a person enters into a different relational space, one which is almost sacred. Whatever the reason for this phenomenon (objective or subjective), those concerned are no longer approached as equals".

The elderly person is one who, being closest to death, comes to represent death among the living: it "enters into history as a living thing", according to Ploton (2004), who also clearly opposes this representation.

Gagnepain subsequently develops two highly relevant ideas. The first is an analogy between childhood and old age, used at various points later in this book:

> "I believe that we should consider the second childhood, i.e. senility, in the same way we regard childhood. Just as, in a normal child, there is no 'child' – the adult is already present – in a normal elderly person, there is no 'elderly', only the eternal".

A healthy child, who already possesses all of his or her faculties on all planes, is, in a way, already adult, and in terms of the anti-psychogenetic logic of the TDM, it is not helpful to see a normal adult or elderly person as anything other than a child who has "become". That said, it is important to distinguish between senility and deterioration resulting from a cerebral lesion, for example. The "normal" elderly person, according to Gagnepain, has not been subject to any change, as a person; in the case of lesions or disease, however, the "person" may have disappeared while the subject remains present. Following this reasoning to its logical conclusion, the author considers that there is no reason to speak of "pediatrics", "geriatrics", etc. On this point, it is doubtful whether Ploton would agree with Gagnepain, as the passage from childhood to adolescence clearly marks the entry into a new phase of life; in this way, it is different from the aging process. Pancrazi-Boyer (1997, p. 50) notes:

> "More generally speaking, we have seen that senescence, like adolescence, represents a period of existential crisis, involving a rebalancing of objectual and narcissic investments. The difference between the two is that with senescence, there is no perspective of 'growing up' at the end of the tunnel, but rather that of the end of the subject. The psychic life of the subject appears to be dependent on the way in which the anxiety connected with the perspective of death is managed".

The author particularly highlights the rebalancing of *topical* investments (interest in stimuli from either interior or external sources) and *libidinal* investments. These two factors provide an explanation for both benign forgetfulness (as opposed to malign, or voluntary, "forgetting") and for the occurrence of what Pancrazi-Boyer poetically refers to as *éclipses de pensée*, "thought eclipses" (1997, p. 51).

The second idea which is relevant for our purposes concerns the difficulty of precisely defining the notion of *aging*. Gagnepain argues that a "theory of aging" is required, as "aging is neither a scientific nor a medical concept". This question is crucial. Although the text is no longer wholly relevant, and significant progress has been made toward establishing memory skill norms for each age range, many specialists – including Brouillet – consider that memory decline in aging patients is not inevitable and that there is a lack of empirical proof. We shall return to this point later.

The subject of Alzheimer's disease is then mentioned, but almost immediately dismissed: "Few things are as poorly-defined as Alzheimer's disease" (Gagnepain 2016, p. 210). Gagnepain introduces the subject by taking an opposite example, i.e. the rare cases where "the person outlives the subject": certain victims of workplace or traffic accidents, suffering damage to the brain, are able to communicate normally despite having lost the capacity for spatial localization. In other terms, the social part of the individual survives but the biological element is altered. In the case of lesions or pathologies, it is more likely that the *person* (social being) will disappear while the biological *subject*[4] remains present. The author's model offers an interesting

4 Gagnepain (2016, p. 140) further clarified this novel distinction, which is essential to his work: "In other terms, above and beyond the being inherent in our nature, i.e. our biology, referred to somatically as the subject, we also possess the capacity to self-construct as a person, emerging as social beings […]".

perspective on Alzheimer's disease, which constitutes a gradual loss on the *person* plane (plane 3). Gagnepain (2016, p. 211) summarizes this as follows:

> "[...] it seems probable that the disappearance of the person is conditioned by a progressive alteration of the cortex, which Alzheimer saw to some degree, but we cannot exactly pinpoint the location. The subject may appear to remain untouched even whilst the person is progressively receding. The patient appears to enter into a form of social sleep, increasingly isolating them from their milieu".

In addition to positioning the disease on the personal plane and in terms of the subject dimension, this definition raises the question of the isolation into which the sufferer descends, little by little. Progressive cortical loss is seen to parallel the loss of the social person. In our speech analyses, we shall see how this theme returns again and again, and is accentuated by an awareness of this "social sleep".

We have endeavored to provide a precise overview of the theory of mediation, which represents the only real attempt at understanding pathologies from a linguistic perspective. However, we shall not go into any greater depth concerning this theory in the rest of our work. We shall simply draw on a number of essential ideas which the theory highlights. First, the way in which deviation from the norm is measured, in terms of linguistic performance, is insufficient: we need to account for patient speech on its own terms, including all of the communication strategies developed by sufferers. The value of this approach is increasingly recognized. Recently, Devevey and Kunz (2013, p. 81), writing on the subject of dysphasic patient care, wrote:

> "Our idea is to consider dysphasic children as thinking, speaking subjects whose variations in usage constitute an idiolect, rather than a language pathology".

Gagnepain avoided this stumbling-block by simply refusing to label speakers who use language in a different way through a pre-established and fixed grid. His model does have its shortcomings, notably in terms of the limitations of the four planes, and the author's comments can also be somewhat obscure; however, overall, Gagnepain's approach is remarkably coherent. The way in which the four planes are mirrored by four registers of language makes it possible to consider a range of different pathologies, justifying the application of the theory as a whole. By observing pathologies (notably aphasia), Gagnepain confirmed Saussure's ideas concerning the bifaciality of linguistic signs and biaxiality, providing a strong argument in support of applicability. He also published work on a number of significant issues which are still fully relevant, such as the way in which Western society views old age (retirement, illnesses, cognitive diminishment) and, more generally, any form of deviation from the norm.

1.2. "Looking for help from linguistics": other (rare) resources

Aphasia was the first subject addressed by the TDM and significantly impacted the way in which it was modeled. It is important to note that this condition is not so distant from Alzheimer's disease; as Fraser *et al.* (2016, p. 408) note:

> "Although memory impairment is the main symptom of AD [Alzheimer's Disease], language impairment can be an important marker. Faber-Langendoen *et al.* [...] found that 36% of mild AD patients and 100% of severe AD patients had aphasia, according to standard aphasia testing protocols".

The fact that aphasia has been a preferred subject for testing other linguistic theories therefore has a logical basis; however, in each case, the authors have focused on finding illustrations to support theories and not the reverse, which makes a significant difference. Hatfield's call for support (1972), "looking for help from linguistics", received very few responses, precisely because of the difficulty inherent in moving from theory to practice. No linguistic model, with the exception of Gagnepain's model described above, has taken pathological speech as its starting point or focused entirely on the subject. The following review is essentially an inventory of the rare attempts which have been made to consider pathologies from a linguistic standpoint.

Jakobson, for example, used aphasia to produce a helpful typological grid, described in detail in *Langage enfantin et aphasie* (Child Language and Aphasia) (Jakobson 1969). The title of this work relates to the similarities between language acquisition in children and the partial language loss inherent in aphasia; however, there are nuances which must be taken into account. First, aphasia does not necessarily involve progressive regression, whereas acquisition clearly follows a regular development pattern. Second, the phonological system used by children should not be seen as the exact opposite of that observed in patients. In his preface, Jakobson (1969, p. 9) notes that the relevance of this analogy is limited to certain cases of aphasia:

> "On closer examination, the language pathology presents a certain level of variation in terms of syndromes, compared to the essentially uniform process of child language initiation. Mirror symmetry between the construction and dismantling of the phonological system is only valid for some of the many and varied forms of aphasia".

Jakobson's linguistic classification of aphasia types was long considered to be authoritative in terms of the specifically linguistic aspect of the condition, and although recent discoveries have led to the introduction of certain nuances, it retains a degree of validity. Jakobson's work intersects with Gagnepain's efforts in several respects: for example, Broca's aphasia is seen as a condition affecting *encoding* and *continuity*, meaning that sufferers lose the capacity for *metonymy*. Wernicke's aphasia is seen to affect *decoding* and *similarity*, thus preventing the sufferer from reasoning through *metaphor*. Martin (1993, p. 10) summarized Jakobson's observations as follows:

> "Jakobson stressed the double operating mode of linguistic systems, with selection (the paradigmatic aspect) operating alongside combination (the syntagmatic aspect). In his view, all aphasic conditions can be defined as a function of these two modes. Similarity disorders affect selection and the choice of units on different levels. Contiguity disorders affect combination and concatenation operations. In aphasia sufferers, the bipolarity between the choice of units and sequential arrangement is eliminated, and only one axis remains operational".

We thus see how, in a certain way, Jakobson was able to use his observations of aphasia in support of pre-established principles, essentially the double articulation between morphemes/phonemes and the opposition between phonetics (sounds, with no distinctive property) and phonology (phonemes with distinctive properties). This may not have been Jakobson's initial intention, given his clear interest for aphasia in particular, but the dominant theoretical orientation is clearly linguistic. Jakobson's work could be seen as a foundation for collaborative work between linguists and neuropsychologists; however, it failed to trigger interest for this area among functional linguists, and no work of this nature – for example, on Alzheimer's disease – was produced as a result.

Nevertheless, Jakobson fills an essential gap, making what we consider to be one of the major contributions to pathology studies by carrying out a contrastive linguistic study. From the outset, the author considered knowledge of languages and their various properties as a prerequisite for studying deviations and regressions in patients; throughout his work, Jakobson provided a number of examples of aphasic patients with different mother tongues.

For our purposes, we have chosen to present a single case which is particularly relevant, as it involves patients who speak Russian, Polish and Czech – three languages of the Slavic family. The heart of the problem relates to vowel length. In Czech, unlike Russian, the use of a short or long "a" sound can be a cause of ambiguities. These are rare when words are used in context, but feature heavily in Czech grammar textbooks. This means that insistence or emphasis cannot be marked by vowel length in Czech, while in Russian, emphasis is typically marked by lengthening vowels. Observation has shown that aphasia can affect this phonological feature in Czech: the patient in question adopted this particular feature of Russian speech. Jakobson stressed that, as each parameter forms part of a system, the loss of the distinctive vowel length characteristic necessarily has an impact on the patient's speech as a whole. In Czech, the first syllable of a word is systematically stressed; the patient shifted stress onto the penultimate syllable, following an accentuation pattern which is standard in Polish, but is only used in Czech for emphasis. Jakobson (1969, p. 106–107) explains this reorganization of the prosodic system as follows:

> "In practice, the expressive stress became the principal stress and took on a configurational role, as the loss of quantity led to a reinforcement of stress. Stress on the penultimate syllable generally appears more pronounced due to the contrast with both the previous and following syllables, which creates a two-sided peak".

The author goes on to explain that this case probably mirrors the prosodic evolution of certain Czech dialects and of Polish. As vowel quantity ceases to be distinctive, prosodic stress is modified. This is not surprising, as any modification to a phonological or morphological system will necessarily have an impact on prosody as a whole. In Old French, for example, the loss of noun cases resulted in the establishment of a fixed component order in the sentence.

Jakobson's principal contribution, aside from the typology of aphasic disorders, was to clearly highlight contrasts between languages, an issue which is rarely taken into account outside of the field of linguistics. A lack of knowledge in this domain, however, can lead to erroneous interpretations of speech characteristics; for example, the speech patterns of the aphasic patient in the example above can be explained by the influence of Polish, but without the relevant knowledge, this influence would not have been detected.

In our examination of psychometric tests, we shall see that linguistic and cultural disparities are generally ignored in the domain of cognitive sciences, despite the best efforts of linguists to raise awareness of the potential issues.

Chomsky and, later, supporters of generative grammar used aphasia to validate the opposition between competence and performance. Whitaker (1968) notably attempted to define whether aphasia affects competence, or on the contrary, should be considered as a disorder affecting the actualization of a given language competence in the context of particular linguistic behaviors. Whitaker (1968, p. 3) rejected the widely held idea that only performance was affected in these cases:

> "It could conceivably be argued that the organic system does not affect linguistic competence, i.e. that aphasia only represents quantitative reduction or impairment in performance and as such is of marginal empirical interest to linguists. The *reductio ad absurbum* of this view is, of course, clear: with maximally severe brain damage there is no evidence of language at all, either externally or internally; under such conditions it seems specious that competence is somehow intact".

In studying patients with aphasia – or other language disorders – linguists are only able to observe performance. Thus, if a competence cannot be identified, it cannot be said to have a neurological substrate. This posed an additional problem for supporters of the generativist model, popular in the 1970s; the notion of empirical validation of the idea of competence had to be abandoned. Even the existence of relevant performance observations cannot be taken as a guarantee of anything. As Ducrot and Todorov (1972, p. 214) noted in the first edition of their *Dictionnaire encyclopédique des sciences du langage*, in the chapter on language pathologies:

> "Two different malfunctions in a set of mechanisms can produce the same type of observable dysfunction at a certain analytical level".

In short, the observed dysfunction is not sufficient to diagnose a specific condition with a sufficient level of certainty. However, in contrast with Ducrot and Todorov's subsequent development, we believe that even a general theory of the normal operation of language would not be sufficient to explain all observed "faults", as identical causes do not always produce identical effects. On the subject of degenerative diseases, many neuropsychologists have noted that almost-identical lesions do not result in the same linguistic symptoms in different patients; it is therefore important to avoid generalizations.

While the functionalist school did not specifically address Alzheimer's disease, the generativist school touched on the condition on several occasions, although the subject was never tackled head-on or as a whole. The central role of syntax in the generativist model makes it particularly well suited to a study of Alzheimer's disease.

Thus, a number of researchers investigated the ability to interpret logical connections between propositions, which require abstract comparison, passive sentences (with agent-patient inversion) and the ability to create and decode anaphoric links. These abilities are seen to be affected in patients with Alzheimer's disease. Generally speaking, syntactic complexity appears to undergo a net regression, as Emery (1988, p. 233) notes:

> "The syntactic data from both the Test for Syntactic Complexity and the Chomsky Test of Syntax converge in the finding that there is an inverse relation between SDAT [Senile Demence of the Alzheimer type] performance and syntactic complexity".

Alzheimer's patients exhibit particularly low scores in all syntactic tests, and for this reason, we shall discuss them further in later chapters. However, this does not signify that these tests are free from bias.

Other authors, such as Kempler *et al.* (1998), have focused on evaluating sentence comprehension. However, *comprehension*, as used in these studies, only relates to the ability to establish correct relations between four types of phrases (simple active, passive, with coordination and with a relative subordinate clause) and illustrations representing these sentences. Kempler challenges the idea that Alzheimer's patients present a genuine syntactic deficiency, and attributes their weak results to memory factors; the author notably argued that syntactic tests overload the working memory, resulting in poorer results in these patients. This type of work suggests that by focusing on patients' syntax, to the exclusion of memory problems, creates a bias in tests. Nevertheless, Kempler failed to apply this reasoning to his own tests, as the images or illustrations used add a further level of interpretative difficulty.

We shall explain later why this extremely common type of test is not particularly relevant in linguistic terms. In the meantime, however, simply note the paucity of linguistic work relating to pathologies, particularly in relation to Alzheimer's disease. While interest has been on the rise since the 2000s, no general linguistic model has yet been found to cover the full scope of language pathologies.

2

Alzheimer's Disease: General Symptoms and Language Impairments

Language impairments in AD [Alzheimer's disease] are characterized by a very high level of variation. There is no single profile which, in and of itself, is able to account for language deficiencies in all patients.

(Barkat-Defradas *et al.* 2008)

The question of language impairments is rarely at the heart of general works on Alzheimer's disease, and discussion is generally limited to a few short paragraphs. The reason for this is clear: following the detection of motor problems, memory alteration and subsequent difficulties in recognition, diagnosis is based on cerebral imaging. Atrophy in the hippocampus is particularly characteristic, as this area of the brain plays a major role in memorization. Nevertheless, as Marsaudon (2008, p. 85) notes:

> "Although the best scanners (spiral, multi-slice, etc.) can now detect cortical and notably hippocampal atrophy, even this examination is not currently sufficient to diagnose Alzheimer's disease".

Any atrophy would need to be shown to be the result of a development – in other words, it must not have been present previously. Similarly, the results of lumbar puncture and biomarker analysis may indicate a likelihood of AD, but are not sufficient to establish a firm and differential diagnosis; they are, however, helpful in eliminating other possible diagnoses, such as strokes, Parkinson's disease, Lewy body dementia, Huntington's chorea or even severe depressive syndromes. Brouillet and Syssau (2005, p. 77) categorically state that "an ATD [Alzheimer-type dementia] diagnosis is a diagnosis by exclusion". Psychometric tests are extremely valuable in this context, as we shall see later. In this chapter, we shall not address risk factors or causes that may lead to Alzheimer's disease; research in this area has not yet produced definitive results. A variety of hypotheses – neuropathological, genetic, vascular, toxic, bacterial, etc. (Brouillet and Syssau 2005, p. 74 *et seq.*) – have been considered, and while these are fascinating, they lie outside the scope of our work.

2.1. General symptoms

It seems fitting to begin this section with a startingly prophetic quotation from Hoffmann, writing in his memoirs (published in 1885[1]) about the construction of the Frankfurt Asylum, of which he was director:

> "The astonishing increase in nervous diseases, particularly of those brain disorders known as psychoses, over the last decades is a matter for concern. Humanity today has become overwhelmingly nervous, and increasingly predisposed to this type of sickness".

Hoffman and his new assistant – a certain Alois Alzheimer – carefully and rigorously implemented their protocol in this "madhouse", identifying and describing the first case of presenile dementia. Alois Alzheimer carried out and reported on frequent interviews, which he combined with post-mortem analyses (and, later, color reproductions) of meticulous, detailed brain cross-sections. Detailed interviews still form part of modern care protocols, used alongside increasingly precise tests. Colored representations

1 Hoffmann, H. (1885). *Lebenserinnerungen* (Memoirs). Insel Verlag, Frankfurt.

of cross-sections taken from histopathological plates, however, have been superseded by modern medical imaging techniques. Nevertheless, modern research is clearly rooted in the work carried out by the earliest "alienists", and Hoffmann was right about the increase in patient numbers.

In this section, language difficulties will be left aside in order to focus on other symptoms. In general, the first observed sign concerns memory loss, which may be noticed by patients themselves or by their family and/or friends. The family physician carries out a first interview, sometimes including a Mini-Mental State Examination (MMSE). Designed to evaluate the patient's cognitive state, this is a series of very general questions relating to the patient's identity and their ability to situate themselves in time and space. This test will be discussed in greater detail later. Patients are then referred to a neuropsychologist, who carries out a more detailed memory assessment using other tests. If the family physician considers that there is a high likelihood of Alzheimer's disease, brain imaging and biomarker analysis may also be conducted. The purpose of the scan is to detect hippocampal atrophy, and biomarker analysis may notably show an anomaly in Tau protein concentration. Finally, retina analysis has been suggested as a way of measuring the extent of amyloid platelets, as the retina is connected to the central nervous system; this would provide a non-invasive means of obtaining early diagnosis.

In addition to strictly clinical data – for which we make no claim of expertise – certain neuropsychologists consider a loss of autonomy, due to increasing agnosia (difficulties in recognition) and apraxia (gestural difficulties), to be decisive. This loss of autonomy is not present in the pre-dementia or Mild Cognitive Impairment (MCI) phase. At this stage, patients have moved beyond the stage of subjective complaints, and their cognitive regression is seen to deviate from the normal aging curve. In short, to cite Hugonet-Diener and Michel (2011) who produced a remarkable graphic rendering of the problem[2], 15 to 20 years may pass between the first signs of

2 We were unable to obtain the rights to reproduce the illustration here. See: "Histoire naturelle de la maladie d'Alzheimer. Progression du déficit cognitif en liaison avec les modifications histopathologiques, biologiques et en imagerie" (adapted from work by Aisen by Hugonot-Diener and Michel, CIPPEG congress, Montpellier, France, 2011).

dementia and patient death; during this period, memory problems will intensify, and may result in a depressive syndrome. At the MCI stage, a reduction in episodic memory performance is already apparent.

As this regression advances, the patient will reach a tipping point, marking the onset of "full" Alzheimer's disease, one to two years before their death. At this point, cognitive loss (detectable using the basic MMS test, discussed at greater length later) reaches a stage where it interferes with social activity, and the patient is no longer able to live independently.

In cases where positive diagnosis is likely, this loss of autonomy is often combined with some degree of unawareness of the problem. Selmès and Derouesné (2009, p. 31) stated that patients "forget that they forget". For a patient to be diagnosed with AD, however, the reduction in autonomy must be correlated with a continuous and progressive alteration of the memory (and vice versa[3]). Here, we have chosen to focus on different types of memory; these types are not all affected in the same way in cases of Alzheimer's disease.

Figure 2.1, taken from Rousseau (2013), provides a general overview of different types of memory.

Long-term memory controls three major processes:

– encoding;

– storing; and

– recall of information.

It is affected from the outset by AD, particularly with regard to the encoding process, which habitually produces new recall cues. As we shall see, Alzheimer's patients do not benefit as much as other patients from an increase in the number of cues used in testing: this indicates a problem with encoding. As Selmès and Derouesné (2009, p. 31) stated, "the recent past is not encoded".

3 "[…] in order to envisage an ATD diagnosis, these memory shortfalls must be observed in conjunction with at least one other cognitive deficit, which may affect instrumental functions (language, praxis, gnosis), executive functions (reasoning, planning, control of execution) or spatio-temporal orientation functions" (Brouillet and Syssau 2005, pp. 77–78).

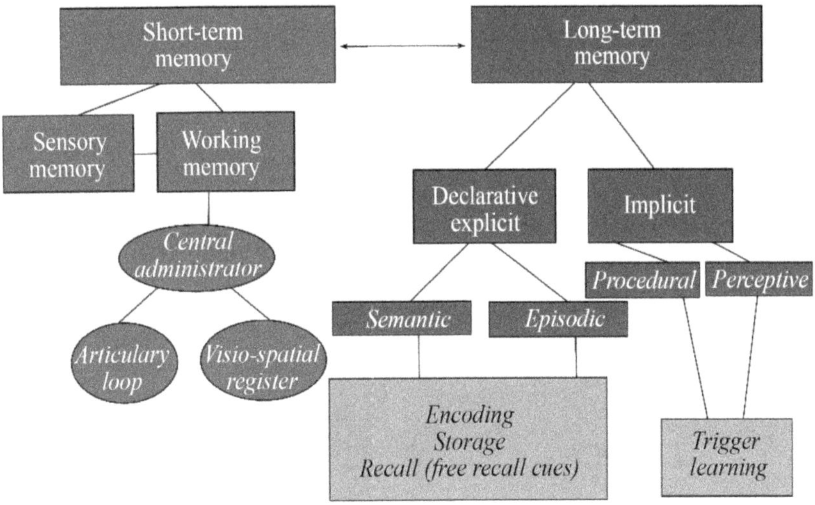

Figure 2.1. *Types of memory (Rousseau 2013, p. 38)*

Different memory types are affected in the following order, although the first two steps can be difficult to separate and may occur concomitantly:

1) *Episodic* memory: "in which a memory or piece of information is associated with a context, notably temporal. This area notably relates to our autobiographic memory" (which concentrates *episodes* of our lives). It is highly emotionally charged, and is thus deeply ingrained 2013, p. 35).

2) *Short-term* memory: Rousseau likens short-term memory to RAM in a computer hard disk. It is also known as *working* memory. This type includes small quantities of temporary data, retained for one minute at most.

3) *Semantic* memory: in this context, the term "semantic" has very little connection to the way it is used in linguistics. It relates to the storage of general knowledge of unknown origin: "This memory provides us with a body of reference, accumulating markers and fundamental information (social and moral rules, rights, responsibilities, etc.) throughout our lives" (Marsaudon 2008, p. 66).

In their remarkable handbook of neuroscience, Bear *et al.* (2007, p. 760) stress the distinction between *declarative* and *non-declarative* (or *implicit*) memories, as shown in Figure 2.2.

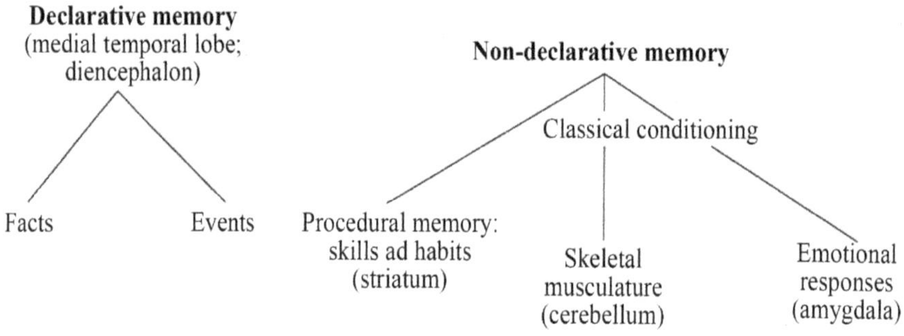

Figure 2.2. *Declarative vs. non-declarative memory (Bear* et al. *2007)*

This distinction is essential as only declarative memory is used for conscious recall; evidently, this is not the case for emotions, for example.

These minor differences in presentation or in the understanding of forms of memory tend to result from a desire to highlight certain elements of data over others, and occur relatively frequently in the literature. For example, a recent science and medicine section in the *Le Monde* newspaper, published on August 23, 2017 and written with assistance from Francis Eustache, a neuropsychologist at INSERM, separates *episodic* and *autobiographical* memory. Box 2.1 is a translation of a box found on page 5 of the article in question, and provides a helpful complement to the diagram in Figure 2.1.

Episodic: the "memory of memories", a long-term memory of events that we ourselves have experienced, situated in a precise spatial and temporal context. This emerges around age 3–5 years.

Semantic: the "memory of knowledge", a long-term memory of concepts, words, general knowledge of the world and of ourselves.

Autobiographical: combines episodic memory with elements from our semantic memory.

Declarative (explicit): concerns the storage and recall of data which we may consciously extract and express.

Procedural: memory of skills and motor, verbal and cognitive abilities which have become automatic.

Implicit: a non-conscious, highly emotive memory. It may control the recall of certain memories, without our being aware of it: for example, it may create a connection between present affects and those of the period in which the memory was acquired.

Perceptive: uses various sensory cues (sights, sounds, etc.) in order to recognize a familiar path, for example, without really paying attention.

Working: a short-term memory that provides us with a "mental workspace", used to store and manipulate information over a period of a few seconds. It thus enables us to complete tasks, carry out reasoning exercises, understand what we are reading, follow a speech or presentation, etc.

Box 2.1. *Definition of different memory types*

There are at least two possibilities for learning and memorization processes, depending on whether or not the sensory information is relayed through the short-term memory (Bear *et al.* 2007, p. 729):

> "Sensory information can be temporarily held in short-term memory, but permanent storage into long-term memory requires consolidation. (a) Information may be consolidated from short-term memory. (b) Alternatively, information processing necessary for consolidation may occur separately from short-term memory".

Consolidation is therefore necessary in order for information to leave a trace. As we shall see later, in the context of tests, the multiplication of recall cues during encoding plays a significant role, one which is all too often ignored.

At this point, it is interesting to look more closely at different forms of memory and at the areas of the brain to which they relate. Brain imaging techniques permit the localization of zones that are activated during different memorization processes. Atrophy of the hippocampus is seen as a strong indication of Alzheimer's disease; Bear *et al.* (2007, p. 744) indicate that:

> "One important role of the medial temporal lobes is in declarative memory processing or consolidation. The best evidence is the severe anterograde amnesia produced by medial temporal lesions".

Many maze-based experiments, mostly carried out on rats (for obvious reasons), confirm the essential role played by the hippocampus in memorization (Bear *et al.* 2007, p. 744–745):

> "If the hippocampus is destroyed before the rat is put in the maze, its performance will differ from normal behavior in an interesting way. In one sense, rats with lesions seem normal; they learn to go down the arms of the maze and eat the food placed at the end of each arm. But unlike normal rats, they never learn to do this efficiently. [...]It appears that the rats can learn the task in the sense that they go down the arms in search of food (the procedural memory). But they cannot seem to remember which arms they've already visited".

Thus, despite an initial appearance of normality, significant difficulties are present – not in terms of traveling through the maze to find food, but in terms of extracting essential information from the recent past. In short, we can state that the rats with hippocampal lesions find food through sheer luck as they move through the maze paths.

Furthermore, if the experiment is modified slightly and food is never placed in certain paths in the maze, these rats will have the ability to avoid the empty paths by integrating old information; however, they will require much longer than the control group to find food. The same reasons are at play here: the "modified" rats cannot remember paths which they have only just taken. As we see, the problem lies in learning new – or the newest – information. In linguistic terms, we would speak of the most *rhematic* sentences or texts. This is why patients often seem lost, anxious or even depressed. They speak in a confused or inappropriate manner: for example, patients may accidentally disclose information that they had previously kept secret, sometimes hurting family or friends in the process. Patients may be or become aggressive, and families are not always prepared for this.

In developmental terms, from the very early stages of the disease, neurofibrillary loss accelerates and communication between neurons is increasingly degraded. Neuron plasticity declines as different possible pathways used to retrieve a given piece of information are cut off. Isnard (2009, p. 84) identified two types of neural lesions that prevent both encoding and recall of older information:

> "A neuron may be damaged in one of two different ways:
>
> – it may die from the inside, attacked by an anomaly in its own protein, which eventually destroys it, having previously hindered its operation along with that of the axon (neurofibrillary degeneration);
>
> – it may be attacked from the outside by an amyloid platelet, a sort of 'glue' that agglomerates around the cell bodies and axons, finally suffocating them".

The conjunction of these two destructive processes constitutes the anatomical marker of Alzheimer's disease, and can be seen in brain cross-sections. Moreover, as Brouillet and Syssau (2005, p. 77) affirm, "post-mortem examination is the only way of establishing a sure retrospective diagnosis of the disease". Although this work is not particularly recent, practitioners still exercise extreme caution. Symptoms of the terminal phase of the disease, immediately before death, are particularly characteristic: without going into detail, "communication is difficult and exclusively non-verbal (language-melodies, touch, looks, etc.)" (Touchon and Portet 2002, p. 46). At the same, severe stage of the disease, other patients are extremely vocal, but speak in a strange, senseless language featuring very minimal syntax and prosody. This representation of Alzheimer-type dementia sufferers reinforces the collective image of these individuals as occupying an intermediary space between life and death, as suggested by Ploton (2004).

2.2. Language-related problems

Moving on from our general presentation, we shall take a closer look at the language-related problems that are central to our work. The order in which we have chosen to present these different elements is itself a subject of discussion, as noted by Emery (2000, p. 146):

> "Although memory impairment generally has been considered the major symptom of DAT [dementia of the Alzheimer type], it is now known that language deficits occur in about 8 percent to 10 percent of Alzheimer patients as a primary symptom early in the disease".

Clearly, language problems must be taken into careful consideration from the outset. This provides an element of justification for the discourse analysis proposed in Chapter 4.

As a sub-set of the cognitive problem group (alongside memory impairments, praxis problems, etc.), language problems can themselves be broken down into different phases and stages of development. Touchon and Portet (2002, p. 39–40) mentioned the following symptoms as characteristics of the first stage, known as the *state* phase:

> "Speech is only weakly informative, poor and may already be partially incoherent. [...] Patients use jargon; periphrases; phonemic paraphrases (e.g. with modification of one or more phonemes in a word) and semantic paraphrases (replacing one word with another, with a similar meaning); neologisms; and perseverations. Echolalia (repetition of a phoneme or group of phonemes pronounced by another) and palilalia (spontaneous repetition of a phoneme or group of phonemes) are also frequent".

As we have seen, diagnosis depends on the fulfillment of a series of criteria (determined through attention, memory and recognition tests) of which language capacities form only one part. In assessments, practitioners pay particular attention to detecting certain specific difficulties (Touchon and Portet 2002, p. 53):

> "The aim is to identify phasic problems, from the simple loss of a correct word to more marked problems in expression, comprehension difficulties, alterations to reading or writing, etc. The analysis of spontaneous speech and of the capacity to understand given directions, naming exercises and verbal fluency tests provide a first indication of these language problems".

However, while Touchon and Portet provide examples of naming tests in the appendix, they do not go into any detail concerning the methods they use to analyze "spontaneous speech", a regrettable omission from a linguistic perspective. Other authors note that, over time, patients have so much trouble remembering words that they use terms only tenuously related to the target (e.g. "nest" for "pigeon") (Barkat-Defradas *et al.* 2008). This feature is particularly characteristic of Alzheimer's disease; while many elderly people exhibit a moderate form of this tendency, AD patients make extremely extensive use of ill-defined words and repetitions, and their use of passive structures and embedded subordinate clauses declines spectacularly (Nicholas *et al.* 1988).

In terms of communication in the strictest sense, we note that, contrary to popular belief, "aphasia, unlike memory impairment, is not a systematically-observed symptom of Alzheimer's disease" (Marsaudon 2008, p. 73). Communication thus appears to be possible, but may be distressing for family and friends due to factors such as forgetfulness and a lack of inhibitions. Patients may not remember things which have just been said and can be peremptory, aggressive on certain topics, or even engage in harassing behaviors. In the most severe stages, mutism is most frequent; however, it is important to consider whether this silence is an effect of the disease itself, or of the isolation that it creates.

Further elements for speech analysis can be found in the work of speech therapists, such as Rousseau (2011). Linguistic parameters (excluding age, socio-cultural background, etc.) are brought to the forefront. Although the author follows the pragmatic theory of language acts, which can be divisive, they note that in a detailed interview setting, oral communication skills can vary enormously depending on:

1) the topic of the discussion;

2) the context of speech (dialog vs. group discussion, differences depending on the person being spoken to).

The language used is also a variable factor: cases of resurgence/forgetting have been observed in bilingual subjects (Barkat-Defradas *et al.* 2008). Use of a foreign language frequently seems to result in a clear modification in patient behavior. Some multilingual patients retain their language skills throughout their entire lives, and continue to use learned foreign languages even at the stage where they no longer recognize their family and friends. Observing a patient whose native language was Italian, we noted that she followed speech much better in Italian, despite having only spoken French for many years. The same phenomenon is evident in Couturier's *Puzzle, journal d'une Alzheimer*. The author (Couturier 2004, p. 112) explains:

> "And then, strangely, I noticed that when I'm faced with people who don't speak the same language as I do, I can follow the majority of conversations. I must pay more attention to their eye movements and gestures. Marlène had already noticed that, because I don't speak a word of Dutch but I was translating words she didn't know for her. That's a bit rich when I can't find an ordinary word in French, my own language, in my own conversations".

In this passage, we note the increased level of concentration involved when a foreign language is used; the author also refers to a feeling, common among sufferers, of having a word in her native language "on the tip of her tongue". The most important element to note here is that any approach to patient speech must involve multiple factors. As we shall see, the tests that are currently used often simply ignore this aspect, perhaps due to a lack of appropriate resources.

As we have already seen, one of the earliest and most widespread symptoms of AD, and a feature contributing to diagnosis, is word loss (Rousseau's "*fuite des mots*", 2013, p. 46), or anomia. It starts out gradually, and sufferers are able to compensate using periphrases; the problem then inevitably grows. The anomia phenomenon accounts for the extent to which studies of Alzheimer's patients focus on vocabulary; in contrast, syntactic problems appear later and appear to be less significant in this type of dementia. Many patients eventually become mute, but others continue to express themselves orally, often using language that is hard to understand. In the latter case, Rousseau (2013, p. 52) notes that "speech is often incoherent (senseless), but sentences are often grammatically and syntactically correct". This supports the idea of an independent syntactic module, which may, as Chomsky suggested, be innate. Furthermore, prosody itself remains intact: this supports the findings of phonologists such as Di Cristo and Rossi, for whom prosody constitutes the very heart of language and not simply a form of "musical" notation. This element may remain after the rest of language is gone, forming the end of a reverse trajectory from full adult capacities to those which we possess at birth, or even *in utero*.

Before going on, it is also helpful to review the chronology of language problems, as determined by Marsaudon (2008, p. 73). The presentation in Box 2.2 provides a helpful summary of the aspects that we have already discussed.

> Beginning: a loss of infrequently used words.
>
> Then: loss of frequent words, replaced by periphrases.
>
> Then: problems with understanding and written expression.
>
> Then: disorganization of written language.
>
> Then: disturbance of oral language, featuring paraphasia[1], jargonophasia[2] and stereotypia[3].
>
> Finally: mutism or quasi-mutism with residual stereotypia.

Box 2.2. *Chronology of language impairments 1 Errors in terms of words and syllables, inversion of similar-sounding words or phonemes. 2 Deformations, confusion, invention of words or phrases. 3 Seemingly uncontrolled repetitions of words or sounds.*

Note that there is some debate over the use of the term "loss", which some authors consider to be erroneous in the context of dementia. Ploton (2004, p. 51), for example, preferred the term *dysfonctionnements* – malfunctions – based on the fact that memories are still present; while it may no longer be possible to access a memory "directly", roundabout routes – such as word associations – can be used to summon up the required information:

> "The patients in question have not truly lost things which appear to be dead or gone. Memories, or capacities for verbal expression, are still present somewhere, but cannot be used in normal circumstances".

The same may be true of language abilities, as Ploton suggests. No explanation has been found for the reappearance of "lost" memories or capacities; however, the existence of this phenomenon means that we must be careful in our choice of terminology. It also offers hope in terms of patient care, even at the level of word "loss", one of the earliest and most characteristic symptoms of the disease.

As we have seen, patients themselves may be the first to note the "loss" of words. Several works have been published on awareness of memory loss in patients themselves; these *metamemory* studies focus on the extent to which patients understand the state of their memory, the processes which they think that they use and their feelings about this state. The data collected in this way is evidently not entirely objective, and must always be considered alongside more objective data in order to establish a likely prognosis for the patients in question.

Taken in isolation, these results pose an unsolvable methodological problem, the same one encountered when using questionnaires on the supposed usage of lexical units in lexicography, for example, or on representations of language (e.g. regional accents) in sociolinguistics. It is therefore essential to use this data alongside scores obtained in psychometric testing, just as corpora are used to provide an additional element of study in linguistics. As Dartigues *et al.* (1997, p. 45) stated: "In conclusion, complaints of memory loss reflect a higher risk of intellectual degradation. They should be taken particularly seriously when communicated to a doctor; nevertheless, test performance remains the best predictor of dementia". However, as we shall see in the following chapter, not all tests are suitable for measuring word "loss", and exhibit significant limitations in reflecting patients' capacity to interpret text.

3

Cognitive Testing: The Key to Diagnosing Memory Pathologies

As we have seen, memory pathologies, such as Alzheimer-type dementia (ATD), are generally diagnosed using a variety of cognitive tests, alongside medical imaging. In this chapter, we shall present a more detailed study of these cognitive tests. In a way, these tests form the key to diagnosis and are used regularly as memory impairments progress.

This third chapter, indispensable in our view, is not intended to provide a complete synthesis of the numerous existing works on psychological tests produced by experts in the field. Our aim is to provide a selective and critical overview of the problems addressed by these expert authors, drawing on the wealth of available literature, and to highlight recurring themes and questions; we do not, however, wish to cast doubt on the status of tests.

3.1. Definitions of psychometric tests

Introductory works on psychometry provide a clear idea of the way in which psychological tests are designed. Anastasi (1988, p. 23) suggests that "a psychological test is essentially an objective and standardized measure of a sample of behavior". The definition given by Eustache *et al.* (2013, p. 50) follows the same outlines, while specifying additional criteria used to define "mental" (more usually called "psychological" or "cognitive") tests; this presentation highlights conditions which must be fulfilled in order to avoid bias:

> "A mental test is a test that is standardized in its instructions, its equipment, the conditions under which it is taken and in the interpretation of the results. It must be reliable – it should be possible to produce the same results, independently of the examiner or moment; it must be valid, providing a specific measurement of what it is supposed to measure; and it must be sensitive, with the capacity to detect individuals who deviate significantly from the norm".

While the adjective "standardized" in both definitions should go without saying, it is important to note the concern for neutrality and objectivity expressed throughout this definition via the notions of reliability, validity and sensitivity. Bernaud (2000, p. 53) stipulates the same requirements, although in a different order: "There are three qualities: sensitivity, reliability and validity". These last two qualities are mentioned by all the authors we consulted (Anastasi 1988; Kline 1993; Jackson 1996; Kaplan and Sacuzzo 2005); the absence of "sensitivity" in the lists given by these authors, most of whom publish in English, is explained by the fact that it is discussed later in their works, in conjunction with test types.

This focus on reliability should be seen in terms of what, in the ill-named "hard" sciences, is known as a "crisis of reproducibility". Both within and outside official and laboratory settings, many have noted the sheer impossibility of reproducing experiments exactly, "all things being equal", and have argued that irreproducibility should not automatically invalidate results; many published and often-cited studies are not, in fact, completely reproducible.

This point is crucial. The notion is not a new one: the *Journal of Irreproducible Results* was founded in Israel in 1955, by Kohn, a virologist, and Lipkin, a physicist. However, the extent of the phenomenon has now become worrying. In 2011, the Bayer company commissioned a group of scientists to reproduce fundamental experiments in various medical fields. The result is edifying: only 21% of the experiments proved to be reproducible in their entirety, and 11% partially. Begley, who was working for Amgen at about the same time, obtained an even lower percentage. In an article co-authored by Ellis and Begley (2012, p. 532), Begley reports, regarding a set of 53 oncology experiments: "It was acknowledged from the outset that some of the data might not hold up, because papers were deliberately selected that described something completely new, such as fresh approaches to targeting cancers or alternative clinical uses for existing therapeutics. Nevertheless, scientific findings were confirmed in only 6 (11%) cases. Even knowing the limitations of preclinical research, this was a shocking result". The response to these results was underwhelming: either the issue was widely recognized but had been swept under the carpet, or the results were considered too embarrassing to publicize further. Continuing down the same path, Begley identified a series of six red flags for suspect work. A negative answer to any of the following questions should alert researchers to non-reproducibility problems:

1) Were experiments performed blinded? If not, the experimenter may have been influenced by their knowledge of the experiment;

2) Were basic experiments repeated?

3) Were all the results presented?

4) Were there positive and negative controls?

5) Were reagents validated?

6) Were statistical tests appropriate and valid?

These six criteria do not provide any real indication, beside neglect or lack of mastery of the complexity of the experiment, why researchers violate them: one reason might be the pressure to publish and be cited quickly. More worryingly, it seems that, in some cases, the reasons why experiments are not reproducible are simply unknown.

Returning to the subject of tests, specifically, all psychometric tests may be subject to the same reliability issues encountered in biological or other experiments. Let us consider two frequent, but no less crucial, examples of possible bias. Firstly, expert neurologists and neuropsychologists have neither the time nor the leisure to administer the tests to patient groups and the control group themselves in order to standardize them. This work is generally carried out by 4th or 5th year psychology, neurology or speech therapy students as part of their master's degree work. While we do not wish to question the quality of testing, we know that a simple difference in instructions can have a visible effect on results.

The second possible bias is a result from the first, generating a chain of consequences: students may have administered the test under different conditions, with different interpretations of the instructions, or may even have chosen to count as wrong some answers which others thought were correct. The pooling of raw data, which is quantitative in nature, then becomes problematic. It is no coincidence that some medical operatives, under cover of anonymity, report that they do not use a particular test because they believe that it contains weaknesses in terms of scaling.

However, we should stress the remarkable uniformity found across introductory literature in the field of psychometry. This is made all the more striking by contrast with the wide range of theoretical approaches to linguistics, for example. Clearly, debates in psychometry do not pertain to the general foundation of the discipline, as a method that is not particularly controversial as it is highly technical and focuses on the valid development of analytical tools. Points of debate are found elsewhere, in relation to phenomena observed in the field – for example on the nature of intelligence and the different types of intelligence which may be identified.

3.2. Test types

Neuropsychology and psychoneurology textbooks (e.g. Eustache *et al.* 2013; Rossi 2013) commonly provide a brief typology of tests, which we shall summarize and discuss below.

The first distinction to be made separates individual tests from group tests, an obvious division which does not call for further comment. A second series of distinctions is suggested by Eustache *et al.* (2013, p. 51):

> "A distinction is conventionally made between three main types of psychometric tests: intellectual efficiency tests, aptitude tests and personality tests, the latter being very rarely used in neuropsychology".

Aptitude tests – which aim to evaluate a subject's capacity to acquire knowledge or process information – are sometimes considered to fall into the efficiency category, along with IQ tests. In this case, only two types of psychometric tests are distinguished. Kaplan and Sacuzzo (2005, p. 9) distinguish between "ability tests" and "personality tests"; the first group is subdivided into "achievement tests" (measuring acquired knowledge), "aptitude tests" (measuring the subject's capacity to acquire a given skill) and "intelligence tests" (measuring the subject's capacity to solve problems, adapt to changing circumstances and draw on their own experience). Jackson (1996, pp. 29–31), following Cronbach (1990), prefers a division into two broad groups: "tests of maximum performance" (with true/false questions) and "tests of typical response" (designed to evaluate behavioral characteristics, without right or wrong answers). While the latter category broadly corresponds to personality tests, the former refers, in a less conventional way, to tests used to evaluate general mental competence (divided into individual or group tests), on the one hand, and to tests aimed at measuring specific mental competences, on the other hand. However, this classification could be seen to correspond to a division between intelligence tests and aptitude tests. Other authors, such as Kline, simply avoid the question. Within Section II (Kline 1993, pp. 171–308), the distribution of Chapters 12 ("intelligence tests"), 13 ("ability tests") and 14 ("aptitude and attainment tests"), alongside the presentation of personality questions from Chapter 15 onward, does not suggest the existence of any particular hierarchy.

Thus, as we see from Table 3.1, there are only a few universal variables. In the latter part of this book, we will focus on individual and efficiency tests (shown in blue in Table 3.1), carried out during memory consultations: these are used to detect specific aptitude impairments, such as apraxia, as well as, and most importantly for our purposes, to diagnose memory disorders.

Individual tests	Group tests
Efficiency (or aptitude) tests	Personality tests
Potentially	
General intelligence tests	Specific aptitude tests

Table 3.1. *Test types*

3.3. Intelligence tests: a starting point

At this juncture, it seems helpful to briefly consider general intelligence tests, as a means of identifying the issues at play in all other test types. These tests – like any other test – raise questions concerning the means and biases involved in their implementation; moreover, as they take us back to the very origins of psychometry, they reveal helpful details pertaining to the initial theoretical objectives of the domain and to biases of tests in general. Authors who mention intelligence tests are duty-bound to cover these points, meaning that epistemological introductions are very helpful for our purposes. The broad outlines of our presentation follow that given by Bernaud (2000) in his short but influential book, *Tests et théories de l'intelligence*.

After briefly recalling the various definitions of intelligence and drawing attention to the fact that tests "infer a level of intelligence from performance in these tests" (Bernaud 2000, p. 11), Bernaud gives a brief presentation of the earliest work carried out in intelligence testing. The original purpose of intelligence (or mental) tests was not to "separate the wheat from the chaff"; Binet, one of the founding fathers of the domain (working in the late 19th Century) aimed to help mentally disabled children. Later, Otis used this approach to optimize the distribution of American soldiers involved in World War I. Interest in extending this approach to other areas (professional and educational) began to emerge after World War II. At this point, the first objections began to be raised, particularly in France, where enthusiasm for the approach was already rather limited (opposition was much weaker in the United States). Bernaud presents the debate surrounding the use of intelligence testing, basing his approach around six broad axes, each featuring two conflicting views. These axes are shown in Table 3.2 (our own interpretation of Bernaud's work).

1	Unitary understanding of intelligence	Pluralist/plural understanding of intelligence
2	Products of intelligence	Processes of intelligence
3	Static understanding of intelligence	Dynamic understanding of intelligence
4	View that intelligence is a socially comparable skill	View that intelligence is a socially acquired skill
5	Intelligence can be measured through abstract tasks	Intelligence can be measured through contextualized tasks
6	Cognitive understanding of intelligence	Cognitivo-emotional understanding of intelligence

Table 3.2. *Different views of intelligence*

While a close study of intelligence tests would not be relevant to our work here, the fact that these tests are the predecessors of all other cognitive tests means that the variables noted by Bernaud remain relevant. The questions raised by these differing understandings are presented below:

1) The debate between proponents of unitary or plural understandings of intelligence is doubtless the oldest. The former argue for the existence of a general factor, *g*, whereas the latter propose a model featuring a variety of different aptitudes. The two views are not, in fact, incompatible; current thinking suggests that a variety of different aptitudes may fort the base of a pyramid with factor *g* at its summit.

2) As a starting point, note that very few tests – of intelligence or otherwise – are designed to evaluate processes. Products – such as test results – form the basis for any inferences of intelligence levels. This approach places the emphasis on predictive value approach, obscuring the means used to solve the problems in question.

3) While intelligence is evolutive in nature and is by no means a static process, testing itself is usually static, i.e. does not involve external help.

4) Bernaud's fourth question is troubling, as it permits observations such as "in the United States, a mean difference of 6 to 7 IQ points has been observed between white and colored subjects" (Bernaud 2000, p. 19). This result is evidently due to outside variables – such as access to cultural facilities – but the fact that such observations may be made is problematic. Bernaud did not see this as a reason to invalidate tests, as "the criterial validity of intelligence tests is broadly comparable from one social group to the next" (Bernaud 2000, p. 19). Nevertheless, one wonders whether, unintentionally, the tests' creators might have chosen items which magnify the impact of external variables.

5) While intelligence often goes hand-in-hand with the notion of adaptation, tests only give partial consideration to this latter element. The capacity for acontextual abstraction is dominant, to the point where the very term "contextualized test" could be seen as an oxymoron. This point will be discussed at length later.

6) This final question is more surprising, and relates to the nature of intelligence itself: it suggests that emotions should also be taken into account, at least in terms of their management. This pathway is an interesting one, but not one which will be explored further here. Note simply that patients who present with memory problems often complete tests in a state of significant emotional distress which cannot be taken into account in scoring.

Clearly, intelligence tests raise major questions; in the context of our work here, the epistemological preference for essentially static and acontextual tests is particularly notable. In order to preserve the full predictive value of intelligence tests, it is essential to make them as objective and authoritative as possible. It is hardly surprising that all published works on psychological (or cognitive) tests and psychometry textbooks have focused on the validity, standardization and mode of administration of tests, alongside potential biases. At this point, the ideological principles of testing are not called into question. Over the course of this work, we shall show that part of the problem with tests of this type lies in the psychometric postulates which they involve.

3.4. The mysterious test

The broad principles of these tests can be found in various accessible works on the question; however, patients report not knowing what to expect when consulting a neuropsychologist, and the tests themselves remain shrouded in mystery. Some patients will have a certain awareness of what is likely to happen, either because they have already undergone testing by a general practitioner (often in the form of an MMSE[1]) or from talking to friends and family; others have no precise expectations and are simply anxious. The difficulty of the tasks and the effort required further increase stress. Some patients raise the objection that no one asks them to complete tasks of this type in real life; as we shall see, this is not entirely wrong. In practice, the neuropsychologist will calmly respond that the reason for the assessment is to test memory *differently*. Medical professionals are trained to deal with this type of defensive attitude, which may persist after testing and can even continue past diagnosis. Certain patients display *anosognosia*[2], or, in other words, are unaware or lack understanding of their memory problems.

---

1 Mini-Mental State Examination.

2 Different explanations for anosognosia have been put forward, notably citing a correlation with emotional impairment: the greater the emotional deficit, the higher the likelihood of anosognosia, as the patient is likely to be less sensitive to the daily effects of forgetfulness. A psychological hypothesis, put forward by Gainotti, was dominant for many years. As Derouesné (1997, p. 60) summarized, "anosognosic subjects are those who, in their lifetime, prized values of independence, prestige and authority. Moreover, anosognosia was more frequently observed in Calvinist Swiss patients than in Italian Catholics".

There is no longer any real secrecy surrounding tests, either on the part of authors or from those who use the tests for diagnostic purposes. Whole works have even been written on the subject, such as the immensely valuable *Gremoire 1* and *2*[3]. In most cases, the authors do not disclose the full contents of tests, but this is essentially for copyright reasons. Nevertheless, the cost of these rights means that those practitioners who use the tests must do so "secretly", using their own translations and without specific training for the test in question – something expressly required by the tests' creators. Other tests, with the appropriate standards and appendices, are freely available on the GRECO (*Groupe de réflexion sur les évaluations cognitives*) website. This particular case is an example of successful cooperation between practitioners and academics, creating tests and the associated standards (evaluation and scoring grids). There are also a number of work groups operating within a hospital setting, and an increasing number of tests feature highly specific, if not *differential*, objectives; some examples specifically aim to identify patients with Parkinson's disease (e.g. the *Mini-Mental Parkinson* test (see Laurent-Mahieux in Hugonot-Diener *et al.* (2015, pp. 15–18)), multiple sclerosis, etc.

Finally, the fact that neuropsychologists are open to new experiences – such as allowing a linguist to observe their work and freely discuss tests with them – also marks a valuable level of open-mindedness.

3.5. Postulates on verbal language in cognitive testing

Before looking at tests in greater detail, it is helpful to establish a clear outline of the problem which we intend to address; we deliberately avoided any precise discussion of this question in previous sections.

Studies on memory diseases are legion, but in our view, none involves a sufficiently ambitious questioning of the postulates guiding the implementation of screening tests and their standardization. Essentially, all or almost all tests on verbal language come from the same theoretical framework, cast from a single mold with a certain number of intertwined

3 The first volume is devoted to tests carried out in the context of Alzheimer's disease and similar syndromes, while the second concerns other neurological diseases with cognitive symptoms. Naturally, *Gremoire 1* is of greatest use for our purposes.

hypotheses as postulates. We shall highlight the issues with these hypotheses in order:

1) words exist in and of themselves, and the mental lexicon is organized hierarchically around prototypes[4], all external to any context of use;

2) to access a word, we thus project the image of the corresponding point of reference, and images of several points of reference may be grouped together according to non-linguistic, acontextual criteria. Furthermore, the word *par excellence* is a noun[5] with a concrete point of reference, and this is sufficient to evaluate the rest of the lexicon;

3) in memory tasks, there is no need to multiply recall cues, as the test is designed to evaluate memory out of context, i.e. a form of "pure" memory;

4) despite the connections between language and culture, cultural differences have little to no impact on the distribution of relations between points of reference/concepts/signified elements;

5) the observer paradox, i.e. the presence of an evaluator, has no significant influence on responses;

6) cognitive capacities, and thus the capacity to use memory, decline with age.

These six tacit postulates result in the idea that it is not only possible but still scientifically admissible to detect memory diseases by measuring the standard deviation between healthy and pathological subjects at different ages in an acontextual manner.

Firstly, we wish to highlight the extent to which modern linguistics, and notably textual semantics, opposes most of these theoretical presumptions, for several reasons. The major objection is that no lexical occurrence, whatever it may be, can be acontextual, even during a neuropsychological examination, which constitutes a context in itself. Working through the postulates in order, we begin by recalling that, in our opinion:

4 For example (Dubois *et al.* 1997, p. 153): "This test consists of learning a list of 16 words from 16 different categories, and is used to assess encoding and selective cued recall. The words to memorize are not prototypes of their categories".

5 Rastier (2001, p. 178) notes that "four-fifths of experiments were based on basic units provided by the analyst: in four out of five cases, these were isolated words, generally nouns".

1) Words do not really have meaning in themselves; meaning always results from semiosis, a unique and temporal pairing of signifier and signified, in a context. This implies that even the signifier does not constitute a natural entry point, and that it must constantly be identified, as work on speech flow segmentation has shown. This fact partially invalidates the semasiological approach; the signified element must constantly be reconstructed, since no two contexts of occurrence are exactly the same. Moreover, prototypes themselves are highly dependent on the context, country, time, etc., and are therefore not systematically identical. These shortcomings – doubtless the result of influences from cognitive psychology, in which language is seen as a structured network of words – are necessarily reflected in experiments conducted on ATD (Alzheimer-type dementia), all of which are designed to test lexical access.

2) Image pairings also depend on the context of enunciation, and it is by no means certain that words with an abstract referent (i.e. non-substantive words) are less suitable for contextualized psychometric use.

3) Furthermore, we believe that it would be helpful to increase the number of recall cues used in encoding, if these are shown to increase effectiveness in patients with certain given memory diseases and not in others. In other terms, if the context of enunciation can be used to help establish a firm diagnosis, it would be helpful to make greater, and better, use of this factor.

4) In addition, we know that cultural differences are marked through the mother tongue, and that the division of concepts operates differently from one language to another. Whether or not we support Sapir and Whorf's suggestion that language is a kind of prism through which we see the world, we can be certain that this division is not identical in all languages; going still further, the naturalness of certain proposed pairings for some speakers might be called into question.

5) Next, it has long been known, in sociolinguistics (see Labov's surveys in particular), that the mere presence of an interviewer[6] affects answers, especially in the case of metalinguistic questions (which concern the speaker's way of speaking and the representations he or she makes of himself or herself) as well

6 Even linguists tend to ignore this factor, as we see from Lee's (2012) highly instructive doctoral thesis, in which this question is never raised. This is not unusual: this bias is also widespread in sociolinguistics, and all studies of "representations" based on questionnaire data present the same weakness.

as in terms of the quality of the answers, which is greatly diminished by the stress resulting from observation.

6) Last, but not least, a decline in cognitive capacity with age is not given. While we make no claim to expertise in this area, note that, according to Brouillet and Syssau (2005, p. 101):

> "Experimental research on aging is explicitly or implicitly based on the 'deficit paradigm'. The results obtained from this research is biased by the simple fact that they aim to confirm the hypothesis that measured cognitive ability declines with age".

In a high-profile article, Rowe and Kahn (1987, p. 143) took a similar approach, stating that: "Research on aging has emphasized losses. In the absence of identifiable pathology, gerontologists and geriatricians have tended to interpret age-associated cognitive and physiologic deficits as age-determined". If we take the time to look, there are many indications that age should not automatically be cited as an explanation for deficits which can be attributed to other causes. In an excellent article, Rotrou and Frambourt (1997, p. 2002) argue that patient habits and practices should be taken into account:

> "These trends in research therefore encourage control, of the 'use-dependency' parameter in studies on ageing; this relates to the amount of practice or degree of exercise of a given cognitive activity. This parameter is more precise than the generalist notion of 'cultural level'. The control of these parameters is particularly relevant when it comes to establishing normative data on cognitive functioning [...], to avoid overestimating the role of age in age-associated decline, i.e. to avoid attributing to age things which are actually due to a lack of practice".

The central idea is illustrated by the following analogy: if an older person is used to skipping rope, testing his or her physical condition through this activity is likely to lead to good results; conversely, a second person who never skips will obtain low scores, which could then be unduly attributed to his or her age.

Furthermore, it seems that seniors' memory test scores are very difficult to arbitrate. As Mahieux and Gély-Nargeot (1997, p. 94) point out: "It should be noted that if we simply construct the Gauss curves of young and old subjects for the tests proposed by Blackford and La Rue, the dispersion of the performances of elderly subjects is much more striking than their actual deterioration. There is a considerable overlap with performances of younger subjects, and the relevance of splitting such curves into slices becomes highly questionable". In light of this surprising observation, and given the scattered nature of their results, it would be unwise to attempt to identify trends in cognitive loss in the elderly. At most, elderly subjects could be divided into three categories: those with no cognitive deterioration, those with cognitive deterioration without dementia and those with dementia. However, it is clear that such a division would tell us nothing about evolutionary trends.

More generally, the opposition between "normal" and "pathological" aging could be reserved for younger subjects, and rejected as an appropriate means of qualifying the elderly. Younger people are presumed to be in good health (cognitive or otherwise), while older subjects all have deficiencies of some kind. Some authors, such as Rowe and Kahn (1987), have proposed a new notion of "successful" versus "usual" aging. This reversal of the traditional understanding leads us to value the absence or relative lack of symptoms, by contrasting it with the usual symptoms of cognitive decline, which, moreover, seems to occur later than we think. Studying the elderly population as a whole, a distinction can be made between extrinsic factors such as level of education, quality of food, etc., and the usual symptoms of cognitive decline. The authors also recommend taking account of the reversible nature of cognitive loss, which they believe is often possible.

In any event, there is clearly no consensus on this issue. This has major consequences for test standardization: how can we identify pathological decline in cognitive ability without a sufficiently precise idea of what constitutes "normal" decline with aging?

Having outlined these issues, we shall now go into greater detail, providing precise examples extracted from actual tests.

3.6. Verbal and non-verbal cognitive tests

Our first impression, when watching patients passing tests, is that the patients' memory is deliberately overloaded until the responses deviate from those of healthy subjects, whether in terms of an extended response time, extensive use of periphrases or the use of hypernyms to replace absent words. This resolutely non-ecological approach has proved effective: deliberate memory overload, punctuated by regular recall cues, is the only known means of detecting memory loss in its earliest stages.

Memory tests can be divided into verbal and non-verbal tests, and this is the approach taken here. As linguists, we have evidently chosen to focus on the verbal branch; we believe that the interpretative process is more interesting and telling when testing is carried out using a text. However, the increased interest goes hand-in-hand with increased difficulty. This is what we intend to discuss in the following sections.

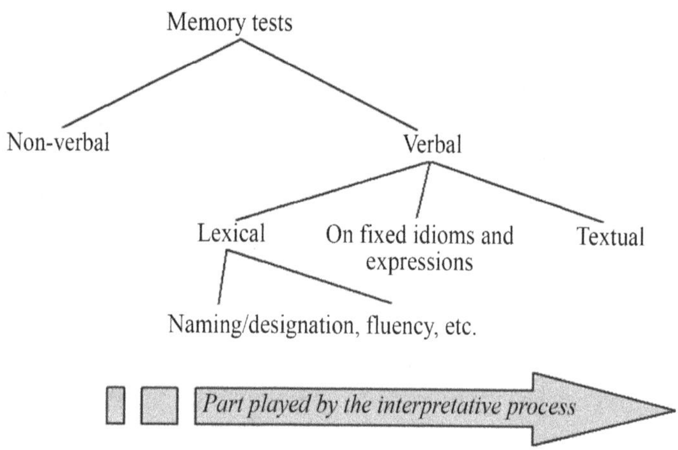

Figure 3.1. *Verbal and non-verbal tests*

3.6.1. *Non-verbal tests*

Although non-verbal tests are of limited interest for our purposes, it is helpful to provide a brief overview of these tests, which form an integral part of the cognitive testing arsenal. Furthermore, it appears that, in some ways, verbal tests are inspired by non-verbal tests, while the reverse is not true. As we have seen, cognitive tests include language as one skill among others; failure in this area results in only one symptom, and difficulties in locating in time and space are also symptomatic of Alzheimer's disease. The presentation given here is not exhaustive, but is intended to lay the foundations for our discussion of verbal tests.

After responding to classic questions on spatial and temporal orientations, patients are asked to reproduce geometric forms such as those in Figure 3.2. In the first case, the pattern often seems to stretch toward the end. The second problem reveals interlocking problems, the third reveals difficulty in assessing spacing and the fourth seems to pose insoluble problems for many patients, who – attempting to copy the shape to the best of their ability – produce a drawing which bears only a distant relationship to the original. The final form often reveals a more or less marked loss of perspective.

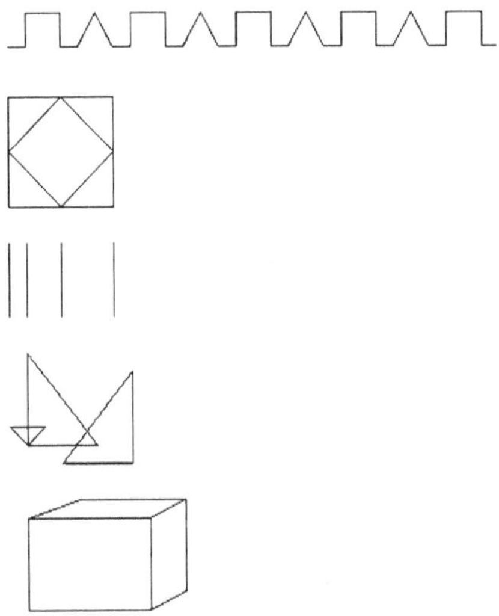

Figure 3.2. *Non-verbal tests 1: copying geometric shapes*

Neuropsychologists also ask patients to carry out a simple search operation, as shown in Figure 3.3, where subjects are required to cross out all instances of the number “5”. The full test features around 10 lines, five of which are shown in Figure 3.3.

	5									
(A)	2	6	5	9	4	5	2	5	2	6
(B)	4	1	2	5	1	3	0	4	9	1
(C)	0	6	7	6	8	9	8	0	8	0
(D)	9	0	4	3	0	1	9	3	7	6
(E)	7	9	5	3	7	8	8	9	7	6

....

Figure 3.3. *Non-verbal tests 2: crossing out numbers*

The test is then made more complex, as the subject is asked to cross out two, then three different numbers. We noticed that these exercises generate considerable stress and that patients make a large number of mistakes; patients seem to think that this basic task needs to be carried out perfectly and quickly. In practice, it is hard to assess whether or not the stress which results from attempting to work quickly is responsible for some of these mistakes. Another number-based exercise consists of storing a series of numbers (in the working memory), which must then be repeated back in the same order, or backward. In the "clock" test, patients are asked to place hands on a numberless clockface to show a given time; in doing so, they must mentally re-place the numbers.

The range of cognitive tests used by neuropsychologists also includes tasks designed to evaluate gestural capacity, as apraxia is often present, alongside other disorders, in patients with ATD. Box 3.1 shows a test in which subjects are asked to reproduce well-known gestures, or to copy manual movements (which may be unusual) made by the evaluator.

Praxis evaluations

Symbolic gestural praxis: "Show me what you do with your hand (or finger) to..."

Give a military salute 0/1

Ask for silence 0/1

Indicate that something smells bad 0/1

Say that someone is mad 0/1

Blow a kiss	0/1	
Total	/5	Normal if > or = 4

Gestural praxis: miming actions

(Pretend to hand over the object): "Show me what I do to…"

Normal = 2/assimilation of body and object = 1/wrong gesture or bimanual ACO = 0

Hammer in a nail	2/1/0	
Tear a sheet of paper in two	2/1/0	
Strike a match	2/1/0	
Comb your hair	2/1/0	
Drink from a glass	2/1/0	
Total	/10	Normal if > or = 8

Abstract gestures

"Do exactly as I do, with your hand"

Right palm on the right cheek	E.M./0/1	
Back of the right hand on the left cheek	E.M./0/1	
Left palm on the left cheek	E.M./0/1	
Back of the left hand on the right cheek	E.M./0/1	
Left hand: horns, II–V	E.M./0/1	
Butterfly	E.M./0/1	
Lozenges II–III (hands reversed)	E.M./0/1	
Double ring	E.M./0/1	
Total	/8	Patient under 65: threshold = 7
		Patient over 65: threshold = 6

Box 3.1. *Gestural evaluation tasks*

The use of the imprecise term "normal" in the scoring is questionable, but the rest does not call for any particular comment and it is not difficult to accept that failure or poor results in these tests indicate the presence of a problem.

Returning to our linguistic standpoint, we will now show that this is not the case for tests that directly involve a language dimension.

3.6.2. *Basic verbal tests: naming, designation and matching*

Within the category of verbal tests, some examples make very little use of interpretative capacities, which are much more important in reading a text, for example. In this section, we shall focus on *naming*, *showing*, *matching* and *semantic fluency*[7] (categorization) tests. We have already touched on the idea that *lexical* tests feature predictable biases, resulting from the neglect of two essential variables: firstly, the *textual* dimension of lexical units, and secondly, their inevitable *cultural* implications. These points will be discussed again here.

As a starting point, let us examine the MoCA (Montreal Cognitive Assessment) test[8], a relatively general examination featuring different types of tasks. The visuospatial/executive section covers material which we have already described, such as the clock test. In practice, the first exercise is highly instructive: patients must connect a series of numbers and letters, where 1 = A, 2 = B, etc. This exercise, as we were able to see in practice, poses significant problems for Alzheimer's sufferers. The second section[9] is a naming exercise, where the patient is simply asked to name what they see. The reverse exercise – where the patient must point to an image corresponding to a pronounced word – is a *designation* test. The MoCA does not feature a *matching* test, which would involve a combination of the first two elements. The example below, taken from Tran *et al.* (2012) (Figure 3.4), shows the three task types.

---

7 The first three tests are discussed in this section, but semantic fluency tests will be left to one side for the moment. These tasks consist of naming members of a semantic category ("animals", "things you can by in a store") for one minute with no repetition.

8 Author's note: I have used the French version of the MoCA, but the terminology used here is taken from the English version.

9 The other tasks will be discussed in greater detail later.

Illustration of the three tasks in the BETL for the target item "glass"

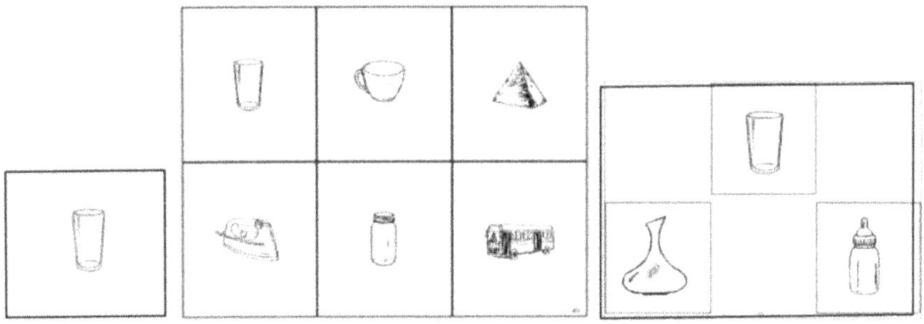

I. Naming II. Designation III. Semantic matching

Figure 3.4. *Naming, designation and semantic matching*

Following the naming test (I), the *matching* test illustrated in parts II and III is explained as follows (Tran *et al.* 2012, p. 1662). *Matching* consists of connecting objects (presented in the form of images) which are considered to be linked in some unstated way:

> "[…] oral lexical comprehension evaluated by a task in which images must be designated in response to an orally-delivered word. Each board features 6 images with, in addition to the target (e.g. […]: *verre* [glass]), a visual distractor (e.g. *bocal* [jar]), a phonological distractor (e.g. *fer* [iron]), a semantic distractor (e.g. *tasse* [cup]), and two neutral distractors (e.g. *pyramide* [pyramid] and *autobus* [bus], one of which corresponds to another target, with no relation to the explored target in terms of form or meaning (in this case, the pyramid). Fifteen boards include one neutral distractor and one mixed distractor, which may be visuo-semantic (e.g. […] *moule* and *huitre* [mussel and oyster]), semantico-phonological (e.g. *oreille* and *orteil* [ear and toe; the French words both look and sound alike]) or even visuo-semantico-phonological (e.g. *escalier* and *escabeau* [stairs and step-ladder])".

Without going into detail for the moment, we see that what passes for a *lexical comprehension* test, both in this example and in the MoCA, is actually a simple matching test with signifiers and signified elements, and at best involves identifying associations. Although this is the intended goal, patients frequently note that the test breaks the laws of communication: a speaker never "naturally" encounters tasks of this type. For subjects who find themselves in a more or less advanced state of language distress, many studies (notably Rousseau (2011)) show that it is essential to avoid jumping around. Although we shall avoid passing judgment on this approach for the time being, this is exactly what happens in these tests: for example, the test following that shown in Figure 3.4 requires the patient to consider a variety of marine animals and shellfish. In her diary, Claude Couturier expresses apprehension concerning certain communication situations, particularly free interviews in which she felt that she no longer had the skills required to participate. She seems more at ease in more "structured" situations, without the prospect of unforeseen events; in linguistic terms, these situations are both *contextualized* and *cotextualized*, forming part of a *textual discourse* and belonging to a *genre*. It is widely accepted that *genre*, as a linguistic variable, creates boundaries in terms of production and expectations in terms of reception. In this sense, we might suggest that these elements be included in lexical tests, especially when working with speakers who are anxious about their situation and who obviously have a good understanding of what is happening to them[10].

Given the history and origins of cognitive psychology (and hence of modern neuropsychology) and of the experimental protocols used in these disciplines, the desire to avoid all contextualization is not surprising. As Rastier (2001, p. 178) points out in an excellent and still very much relevant book[11]:

> "[...] neurolinguistics and psycholinguistics share the same preconceived notions of language and meaning used in AI [Artificial Intelligence] and elsewhere in cognitive research, notably the reduction of linguistics to the verbal (sound) aspect; the separation of signifier and signified promotes a modular understanding of language, assimilating signified elements to representations".

---

10 As Guertin (2000, p. 85) notes, speaking on the subject of tests, "in the eyes of the relevant authorities, the level of the patient may have been assessed correctly; from the patient's perspective, however, he or she has been accused of failing the test, told they are no good, in short, has been deemed worthless".

11 We refer to the 2001 edition of *Sémantique et recherches cognitives*.

Considering these remarks, it becomes clear why words with concrete referentials are always used in testing. At this point, we can formulate a number of objections to these tests:

1) These tests only use nouns with a concrete referential, which certain speakers may rarely, if ever, have encountered (take the case of marine animals, for example, which a patient may never have seen), which implies that these terms are unlikely to have been used on a regular basis.

2) The tests do not take sufficient account of differences in the vocabulary skills of different patients before diagnosis and observation; note that the maximum level of academic study is not sufficient to assess this level (e.g. a university graduate will not necessarily have a more extensive vocabulary than an expert Scrabble player with less formal education).

3) Furthermore, they ignore the construction of semantic fields, as defined in corpus linguistics. We know – from numerical and statistical data – that there are attraction/repulsion trends between lexemes which are not necessarily obvious to non-linguists, and practitioners cannot identify these in advance. For example, as Rastier demonstrated, "caviar" and "fish sticks", both of which possess the /food/ seme, might be expected to occur together frequently in context; however, corpus studies have shown this to be extremely rare (even in recipes), and the secondary /luxury/ seme for "caviar" acts as a barrier between the two[12]. A list based on actual corpus analysis could result in connections between *champagne*, *caviar, foie gras*, *canapés, salmon* (*smoked* or otherwise), *apple* and *truffle*; inversely, *fish croquettes* or *orange juice* would be excluded from a list of this type.

In reality, words are chosen on the basis of *frequency*, *length* (despite the fact that the *density of the phonological neighborhood* seems to be more important for word retrieval) and the opposition between *manufactured* and *natural* objects. The 60-item Boston Naming Test (BNT), which, with its

12 "In a speech situation, caviar is much more likely to be found in the vicinity of champagne or *foie gras* than of fish sticks which, theoretically, according to the WordNet network, is only one hop away!" (Rastier 2011, p. 154). The two could be expected to coincide in a recipe corpus, but in reality (thankfully) this has never been attested.

numerous variants, is something of a benchmark in neuropsychology, involves representations[13] of the following elements:

> 1. Bed. 2. Tree. 3. Pencil. 4. House. 5. Whistle. 6. Scissors. 7. Comb. 8. Flower. 9. Saw. 10. Toothbrush. 11. Helicopter. 12. Broom. 13. Octopus. 14. Mushroom. 15. Hanger. 16. Wheelchair. 17. Camel. 18. Mask. 19. Pretzel. 20. Bench. 21. Racquet. 22. Snail. 23. Volcano. 24. Seahorse. 25. Dart. 26. Canoe. 27. Globe. 28. Wreath. 29. Beaver. 30. Harmonica. 31. Rhinoceros. 32. Acorn. 33. Igloo. 34. Stilts. 35. Dominoes. 36. Cactus. 37. Escalator. 38. Harp. 39. Hammock. 40. Knocker. 41. Pelican. 42. Stethoscope. 43. Pyramid. 44. Muzzle. 45. Unicorn. 46. Funnel. 47. Accordion. 48. Noose. 49. Asparagus. 50. Compass. 51. Latch. 52. Tripod. 53. Scroll. 54. Tongs. 55. Sphinx. 56. Yoke. 57. Trellis. 58. Palette. 59. Protractor. 60. Abacus.

Clearly, the subject's memory is projected from taxeme to taxeme with no semantic indications. Note that many lexicological studies, particularly in polysemy, are subject to the same issue, artificially grouping accepted meanings which do not share a context of appearance. This comment is relevant as certain speech therapy tests feature this theoretical error. As professionals such as Castera *et al.* (2010) have remarked: "The selection of words is developed following work on polysemy, whereby different meanings of the same word are identified using a contextual cue". The authors cite the example of the French word *Baguette*, which has multiple context-dependent meanings. For a baker, this is most likely to be bread; for an orchestral conductor, the *baguette* is the baton; and for a carpenter, a *baguette* is a length of wood doweling.

It is very unlikely for these multiple meanings of "baguette" to be found together in a single context (unless the speaker is joking), meaning that speakers with memory problems will need to use artificially connected lexical networks. In this case, the baker's *baguette* belongs to the /food/ category, while the conductor's *baguette* is a /tool/.

---

13 The illustrations are not always particularly good. Furthermore, practitioners only rarely have access to an original set, in color, and work from black-and-white photocopies.

To be more instructive, the tests could be contextualized by proposing only images of referents, possibly divided into cycles, which are likely to appear in the same context. In short, the most problematic aspect is that existing tests violate the hermeneutic law of determining the local by the global, which increases the affinities between certain lexemes and drastically reduces the chances of seeing others side by side in a given text.

Yet, these constraints constitute an essential part of language in the way it is learned, internalized and rendered by speakers. For example, it is astonishing that Spinelli and Ferrand (2005, p. 164), when faced with speakers' responses for "the first word that comes to mind after being presented with a stimulus word" (in this specific case, *mouton* [sheep][14]), should have been surprised (or, at best, feigned surprise) at the wide range of different answers. As we see from Figure 3.5, *sheep* generated a range of answers which cannot all be explained through neurolinguistics: the "associative links" (to use the authors' term) at play are conditioned by culture and social practices, which are essentially left aside in the field.

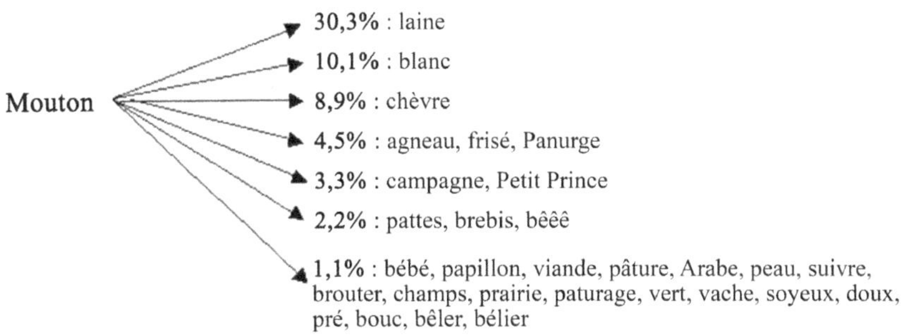

Figure 3.5. *Responses to the "mouton" stimulus. In English, top to bottom and left to right: wool, white, goat, lamb, curly, Panurge, countryside, Le Petit Prince, hooves, ewe, baa, baby, butterfly, meat, pasture, Arab, skin, to follow, to graze, fields, prairie, pastures, green, cow, silky, soft, meadow, billy goat, to baa, ram*

We should add that this study created as much work for its authors as to the respondents, forcing them to make arbitrary groupings (e.g. *campagne*, "countryside", with Saint-Exupéry's *Le Petit Prince*). The results of such

14 A noun, evidently, due to "the dominant position of nouns in language philosophy, where this morphological class is traditionally used as a medium for ontological questioning" (Rastier 2001, p. 178).

exercises could even be said to contradict the postulates and methodology which underpin them, without the neuroscientists who produce them being aware of it. Language is not a nomenclature, and words are not unilaterally linked to referents, nor are individual words associated acontextually. Interpretation is first determined by social practices and culture, then by corpus and genre, and finally by text and syntagma. This test, like so many others, uses a word artificially taken out of context; as it does not provide the set of definitions required for interpretation, its results are not generally applicable. The authors are therefore trapped; at best, they may simply conclude that, in an artificial communication situation such as a psycholinguistic survey, 30% of French speakers consider /pourvu de laine/ (woolly) as the most prototypical feature of *mouton*[15]. We also note that the tests described above make intensive use of the term "stimulus", a word inherited from behaviorist theories, which should be considered in conjunction with subjects' responses.

To eliminate some of these methodological biases in experimental protocols, practitioners could propose short texts (or short excerpts from texts within the same genre) containing images that represent objects to be named, as in some children's books, or at the very least, images of referents that are statistically more likely to occur together in a corpus. This would ensure thematic continuity. This option does not appear in the two books published by GRECO, and this omission is telling. Texts, and the associated cultural and social dimensions of the interpretative process, are surprisingly absent from cognitive testing. The 3.3% of subjects for whom the word "sheep" is most closely linked to *Le Petit Prince* lose a valuable point, as do the two out of three subjects who say "Bambi" in response to a picture of a doe.

3.6.3. *Tests using idiomatic expressions*

Some tests go beyond the level of words alone, assessing patients' understanding of fixed expressions. These are notably used in cases where a patient is suspected to have early-stage Alzheimer's disease.

15 It is hard to eliminate the possibility that other, more instinctive responses to the stimulus, such as *bêêê* (baa), are an effect of context.

As we aim to demonstrate, these tests take a similar approach to those described above, and feature the same biases. They differ in that the linguistic units used are more complex, featuring fixed (non-compositional) expressions with a possible compositional interpretation. For example, the phrase "throw in the towel" could be interpreted literally, by breaking it down into two principal lexemes, or figuratively, as a synonym of "abandon". It is relatively simple to design experiments based around the existence of two interpretative pathways; furthermore, "Findings suggest that dementing patients have problems particularly in complex linguistic operations like proverb interpretation" (Code and Lodge 1987, p. 34). Other authors, such as Kempler, have gone still further: they suggest that one of the most obvious linguistic markers of Alzheimer's disease is a loss of mastery of abstract language. If they only possess a literal understanding of language, Alzheimer's patients may feel insulted if they interpret an abstract expression literally. The capacity for abstraction is often assessed using fixed expressions, based on the literal/figurative opposition; this has been the subject of considerable debate since the 1980s, and no consensus has yet been reached.

Leyhe *et al.* (2011, pp. 51–61) present an excellent example of this approach. The authors aimed to measure the abstraction capacity of patients presumed to have early-stage Alzheimer's disease. To do this, they presented subjects with a list of 10 proverbs – e.g. "While the cat's away, the mice will play" – which they were asked to interpret. A scoring chart was developed based on various different interpretations, featuring five different types, organized along two axes, one binary (meaningful/senseless) and one scalar (abstract/partially concrete/concrete), which the authors referred to as the "degree of symbolization". This scoring chart is shown in Table 3.3.

The test featured nine further proverbs (originally in German), such as "too many cooks spoil the broth", etc. The results of this task, carried out alongside others of the type described earlier, seem to confirm a decline in patients' capacity to master abstract thought. One might object that the capacity for abstraction is measured here using a socially and culturally normative meaning for a given expression: patients are not asked to think *per se*, but simply to reproduce the commonly accepted meaning of a fixed expression.

		> Desymbolization >		
		Abstract	**Partially concrete**	**Concrete**
Sense	**Meaningful**	Type I: when no controlling person is present, you can do what you want	Type II: when nobody is there to look after you, the mice can do what they want	Type III: cats eat mice. Mice hence cannot celebrate till the cat has left home
	Senseless	Type V: when nobody is there, you can do anything alone	No type available	Type IV: when the cat and the mice are absent, then the house is completely unattended

Table 3.3. *Desymbolization of idiomatic expressions, illustrated for "While the cat's away, the mice will play"*

Other authors working on this subject, notably in the United States, have focused on the suppression process. This interpretative operation, known in semantics as *inhibition* (which Rastier contrasts with the *activation* of semes) or *narcotization* (Eco), consists of retaining only those semes which are relevant in a given context. In the case of "throw in the towel", idiomatic interpretation would be made possible by assigning meaning to the expression as a whole but not to its components: in this way, all of the denotative semes for "throw" and "towel" are inhibited. The division depends on the context, but the two interpretations are not strictly speaking mutually exclusive and do not in any way enter into complementary distribution. In many cases, there is even a *syllepsis* of meaning, i.e. a situation of semantic duplicity, often used by advertisers, comedians and satirical journalists. For example, when a transport firm produces an advertisement featuring a picture of a train and the slogan "we go a long way for you", both interpretative pathways for "go a long way" are activated simultaneously by the visual.

This clearly demonstrates why many authors consider that the literal sense of components remains active even after the full fixed expression has been seen, and even if the context would suggest a figurative interpretation (see Hillert and Swinney (2001) or Colombo (1993)): as Cieślicka (2011, p. 16) notes, "The activation of idiom literal and idiomatic meanings is obligatory, irrespective of whether context favors its literal or idiomatic interpretation". Of course, fixed expressions are not all created equal: the meaning of some is more evident than others, and for some of these expressions, a literal interpretation, in referential terms, would be difficult if not impossible.

A more elaborate version of Leyhe's test involves a visual element, used to exclude the literal meaning of the expression. This approach, known as sentence-to-picture matching, presents certain similarities with simple lexeme naming tests.

Idiomatic expression

[Distractor] reformulation of literal sense no. 1

[Correct answer] reformulation of idiomatic sense

[Distractor] reformulation of literal sense no. 2

Figure 3.6. *Sentence-to-picture matching*

Patients are shown a visual representation of the idiomatic interpretation of an expression and are asked to select from three responses: one "correct" and two "distractors" which are simply reformulations of the literal interpretation. The problems relating to a lack of context, identified in previous tests, are still relevant here: the sole context provided in this case is a visual representation, and it is very difficult to assess the relevance and accuracy of these representations for all subjects, sick or otherwise. Nevertheless, following Andree *et al.* (1992, p. 33), it seems that in the earliest stages of Alzheimer's disease, patients still understand idiomatic expressions such as proverbs, even out of context. This ceases to be true in later stages, where the visual context becomes essential to understanding. On the other hand, many neuroscientists have noted that tests of this type involve too many variables, making simultaneous demands on the memory, attention, concentration and vision (with recognition), all of which combine with sociocultural factors and, seemingly, the level of education.

There are two points to note with regard to these tests, additional to our previous observations:

1) Regarding the minimalist view of the context of use of lexia/ lexemes/idioms: as in the case of naming tests, a simple image is used to match signifier with signified in a referential approach. As we have seen, this is problematic due to the multiplication of variables and, *in fine*, makes it difficult to isolate an inhibition of the idiomatic sense.

2) Regarding the assessment of "abstract" thought via rememorization and (re)production of an absolutely normative sense of specific linguistic units, i.e. fixed expressions and proverbs. One wonders if fixed elements and proverbs are entirely suitable for use in assessing patients' capacity for abstraction, due to their non-compositional nature. If the meaning of the selected units cannot be "calculated" from their components, then correct interpretation depends more on memory than on abstract thought.

3.6.4. *Barbizet's test: "The Lion's Tale"*

To the best of our knowledge, only a few tests have challenged this hegemony of the purely lexical level. Tapiero presents research based on subjects recalling a "tale", but this "tale" is extremely short (just four sentences)[16]. The best-known test using a longer text is Barbizet's "The Lion's Tale", developed in the 1960s, at a time when the use of theoretical paradigms in experimentation was much more limited. As we shall attempt to show, although this test is strictly textual, it also highlights new biases. It has fallen from use in recent years; a lack of standardization and effective weighting mean that it does not produce easily exploitable data, except for patients who are already in the later stages of ATD.

The test consists of reading a short story of just under 100 words, once at the start of the experiment, then again 15 minutes later. Barbizet and Truscelli's original plan was to assess patients' memory of the tale one hour,

16 Cited by Rossi (2013, p. 57), although not referenced in the bibliography: "One day, a giant walked through a forest. The giant saw an old lady. The old lady was collecting dead wood. The giant decided to help her" (translated into English by the author).

one day and even one week later, but this is hard to implement outside of a hospital context. However, the test is no less interesting when used alone in the context of memory consultations. The use of delayed recall is used to measure *confabulation* – this was, in fact, the original main purpose of the test – as well as to assess whether the length of the text and the introduction of a story are relevant. The Lion's Tale was designed to identify impairments resulting from dementia, as well as from strokes, tumors and even alcoholism. It can therefore be used in the context of Alzheimer's disease, which is our main focus here. Furthermore, according to Croisile, in *Gremoire 1* (Hugonot-Diener *et al.* 2009, p. 111), Alzheimer's sufferers systematically score lowest on the test, both for immediate and delayed recall, showing the highest levels of forgetting. Simple anxiety or depression, on the other hand, will not lead to a higher than normal level of forgetting, except in situations of complete blocking.

The proposed text is as follows:

> "Un lion / nommé Sultan / s'échappa de sa cage, / par la porte mal fermée / par un gardien négligent. / La foule des visiteurs / nombreuse ce dimanche-là / s'enfuit vers les bâtiments voisins. / Une femme, / vêtue de bleu / qui tenait dans ses bras son enfant / de un an / le laissa tomber. / Le lion s'en saisit. / La femme en larmes / revint sur ses pas, / et supplia le lion de lui rendre son petit. / L'animal la regarda / longuement, / fixement, / et enfin lâcha l'enfant / sans lui avoir fait le moindre mal"[17].

Neuroscientists use a copy of the text, broken down into "components" (mostly nominal, verbal or adverbial groups), each on a different line. Each component has the same coefficient, 1/22, as shown in Table 3.4.

17 Author's note: the test was conducted in French. As the author notes, language and culture are not neutral; we have thus refrained from translating the tale and elements of the tale within the body of the text. An English translation follows, for information purposes: "A lion / named Sultan / escaped from his cage / through a door which was left open / by a careless guard. / The crowd of visitors / who were numerous that Sunday / fled to the neighboring buildings. / A woman / dressed in blue / who was carrying her child / aged one / dropped him. / The lion took him. / The woman, in tears / turned back / and begged the lion to give the child back. / The animal looked at her / for a long time / intently/ and finally let go of the child / having done him no harm".

Number	Propositions	Immediate recall	Delayed recall (15')
1	Un lion		
2	Nommé Sultan,		
3	S'échappa de sa cage,		
4	Par la porte mal fermée		
5	Par un gardien négligent.		
6	La foule des visiteurs		
7	Nombreuse ce dimanche-là,		
8	S'enfuit vers les bâtiments voisins.		
9	Une femme,		
10	Vêtue de bleu		
11	Qui tenait dans ses bras son enfant		
12	De un an		
13	Le laissa tomber.		
14	Le lion s'en saisit.		
15	La femme en larmes		
16	Revint sur ses pas,		
17	Et supplia le lion de lui rendre son petit.		
18	L'animal la regarda		
19	Longuement,		
20	Fixement,		
21	Et enfin lâcha l'enfant		
22	Sans lui avoir fait le moindre mal.		
Total		/22	/22
Errors and confabulations			

Table 3.4. *Propositions in Barbizet's "The Lion's Tale" test*

The instructions seem clear, at first glance: the practitioner asks the patient to "remember the general outlines of the story" and/or "remember as many details as possible". While the second part of this instruction is acceptable in a clinical detection context, it runs contrary to the way a speaker would usually interpret a text: in a "natural" situation, an individual would never attempt to remember everything or nearly everything, but would "sort" the information. Not every element is equally important; subjects will pay less attention to descriptive details which are not essential to the narrative. Croisile (Hugonot-Diener *et al.* 2009, p. 109) suggests that the most salient items are nos. 1, 3, 6, 9, 11, 14, 17, 21 and 22. However, the soring table for the test is binary: each element is weighted equally, and is either right or wrong. Neuropsychologists also have a certain degree of freedom in interpreting the results, as "certain words are strategic, and their absence or substitution removes the point of the proposition" (Hugonot-Diener *et al.* 2009, p. 110). These "strategic" elements are those which are crucial to the narrative, i.e. which relate to the *process* and the *participants* involved in the story – the *petit drame*, as Lucien Tesnière put it. Adverbs (*fixement*) and their paradigmatic equivalents, including most of the prepositional constructions (*par la porte mal fermée*) and noun expansions (*de un an*), are logically excluded from this category. Clearly, the fact that the lion caught the child – *le lion s'en saisit* – is of far greater narrative importance than the fact that the woman was wearing blue – *vêtue de bleu.*

Even those aspects which initially appear clear could be debated and improved upon. Croisile's separation of salient/non-salient propositions is arbitrary and subjective in nature. In French, the most rhematic elements – those which relay the newest information – are generally situated at the end of a sentence in French, when the SVO (subject–verb–object) syntax is present. In this case, only elements at the start of sentences (propositions 1 through 5) are considered salient, which excludes the circumstantial information given at the end of the phrase. Strangely, in the following sentence (propositions 6 through 9), the verb *s'enfuit* is not considered salient, unlike the subject, *la foule des visiteurs*. It should, perhaps, have been included to maintain the parallel with the first sentence.

Thus, we see that *la foule des visiteurs* is a low-information component, and the predicate which accompanies it should not be considered important. All things considered, the informational weight of syntagma appears to be scalar and cannot reasonably be evaluated using a Boolean approach.

Let us start with our first remark: if, as it seems to us, not all information belongs to the same level, then elements should be weighted as a function of informational value, using a scale. This has already been noted by certain authors, including Rossi (2013, p. 57):

> "It is also possible to vary the weight of recalled information as a function of criteria defined by the researcher. Thus, macrostructural information (i.e. which is structurally important) might be scored using a higher coefficient than microstructural information (details)".

To respond to the second issue, each element could be assigned an informational value from 1 to 3, giving evaluators the option to give one point out of two or three, for example, for imperfect or partial but coherent responses.

To do this, we could use syntactic functions – as described above – as well as, and better still, *referent strings*. Looking more closely at the text, we see that:

– lines 1–13, while describing actions, also constitute the *referent introduction* phase;

– lines 1–5 introduce the "lion" referent (and, in passing, the guard – "gardien" – who is not mentioned again);

– lines 6–8 introduce the crowd, "la foule", then by extraction of this *hypertheme*, the sub-theme of the woman, "femme", and her child, "enfant", who remains in a *patient* role throughout the text. This gives us a *derived theme progression*, as identified by Combettes.

The remainder of the text corresponds to the *referent reactivation* phase, where both the characters and information previously given concerning them are used. We encounter brief action elements, as in line 14, "le lion s'en[18] saisit", followed by a negotiation, where the woman asks the lion to give back her child. Patients often appear to forget about the negotiation, simply remembering that the mother begs the lion to give the child back, and that the child is given back to its mother. The final line is crucial, as, in addition to concluding the tale, it contains information of high rhematic importance. Just imagine if the child had been bitten by the lion… This conclusion should never be omitted by patients, and the highest possible score should be

18 Note the use of the adverbial pronoun "en", which is rarely used in practice to refer to an animate object, let alone a person. This usage could potentially lead to a false interpretation (although the only possible point of reference here is the child), or at least to a brief hesitation.

used for this final memory trace. In summary, the text features three essential referent strings:

– the *lion* "nommé Sultan" – named Sultan – who essentially completes four actions: "s'échappa", "s'en saisit", "la regarda" and "lâcha" (escaping/catching the child/looking at the woman/dropping the child);

– the *femme* (derived from the *foule*) "qui tenait son enfant", "le laissa tomber" and "supplia" (was holding her child/dropped the child/pleaded);

– the *enfant*, who, as we have said, simply undergoes actions, acting as an object complement.

We might also consider that the identity of the three protagonists should absolutely be retained, perhaps with a degree of tolerance for confusing the child, aged 1 (the noun and the expansion of the noun are spread over two lines, 11 and 12, for no apparent reason) with a "baby", a confusion we were able to observe in practice. All three of these elements, along with the conclusion in line 22, might be assigned three points. Each action could also be worth three points. Next, patients are likely to remember the name of the lion, "Sultan", for two reasons. Firstly, the woman has no name; secondly, "Sultan" benefits from the primacy effect, well-known in neuropsychology, which means that patients are more likely to remember the first elements mentioned. This element, along with the initial cause of the drama, "la porte mal fermée", "la foule qui s'ensuit" and a number of other secondary information elements, could be given a coefficient of 2. All remaining elements would be assigned the lowest coefficient, 1.

Based on these recommendations, Table 3.4 could be modified as follows. Firstly, information which can be omitted without loss of points is placed in brackets. When information is missing, the total number of points for the line (between 1 and 3) is lost, unless the evaluator considers that the patient's response was partially correct. While this option is eminently subjective, note that it is impossible to remove all subjectivity from an interpretative activity; note also that the final line of the table is intended for recording "errors and confabulations" by the patient, and that no one has exactly the same idea of how to evaluate these. Barbizet provides no further information on this subject.

Number	Propositions		Coefficient	Immediate recall
1	Un lion		3	
2	Nommé Sultan,		2	
3	S'échappa (de sa cage),		3	
4	Par la porte mal fermée		2	
5	Par un gardien (négligent).		1	
6	La foule (des visiteurs)		2	
7	Nombreuse ce dimanche-là,		1	
8	S'enfuit (vers les bâtiments voisins).		3	
9	Une femme,		3	
10	Vêtue de bleu		1	
11	Qui tenait dans ses bras		3	
12	Son enfant de un an		3	
13	Le laissa tomber.		3	
14	(Le lion) s'en saisit.		3	
15	La femme en larmes		1	
16	Revint (sur ses pas),		2	
17	Et supplia le lion (de lui rendre son petit).		3	
18	(L'animal) la regarda		3	
19	Longuement,		1	
20	Fixement,		1	
21	Et enfin lâcha l'enfant		3	
22	Sans lui avoir fait le moindre mal.		3	
Total			**/50**	**/50**
Errors and confabulations				
Legend	*Themes*	*Predicates*	*Secondary information*	

Table 3.5. *Proposed improvements to the scoring grid*

Table 3.5 also features different shades of gray, used to mark different types of information (themes, predicates and secondary information). These distinctions do not affect the coefficient (a theme is not automatically assigned a value of 3, etc.), but this approach should make it easier to detect patterns of forgetting. Note, too, that the total score of 50 is much easier to convert to a percentage than the original total of 22.

With our modifications, sorting elements by type, the test could even be used to create patient profiles for different pathologies. Currently, the test is only used to confirm and refine impressions obtained from other tests[19]. Instead, it could be used systematically as part of detection testing. It has the notable advantage of being more "fun", as a narrative, than a series of images to name or connect. As we have attempted to show, it is also adjustable, and could be used to establish a typology of forgetting which is potentially significant.

Furthermore, the first results obtained using this version of the test[20] seem to reveal a trend: patients with suspected Alzheimer's disease systematically remember the beginning and the end: in short, the essence of the tale and its conclusion are all that remain once the rest has been forgotten. While the results obtained using Croisile's scoring system and our own system are, overall, comparable, as we expected, a closer look at the details of the test reveals a number of interesting elements. Generally, the most salient items (those with a coefficient of 3) are remembered, but those with a coefficient of 1 are forgotten, except in rare cases where the overall score is low. Furthermore, patients seem to have a clear propensity to remember predicates, more than themes, and considerably more than secondary information. Finally, confabulations are observed in patients with the lowest MMS.

We shall return to the subject of confabulations later, but for now, we wish to draw attention to the importance of instructions; differences in this regard can have significant effects on the results, and the hypothesis that responses may have been affected by ambiguous directions cannot be

19 "Barbizet's Lion's Tale test confirmed the presence of a memory disorder, with only 7 of 22 items recalled immediately" (Poujois *et al.* 2010, p. 29).

20 At the time of going to press, this test is being used by masters' students in speech therapy school, who aim to measure the difference between results obtained using this version of the test and using Croisile's version of the test.

eliminated. Telling a patient to attempt to remember everything (which is relevant from a psychometric standpoint, but not particularly "natural") is very different to telling them to remember broad outlines (in a natural, but hardly quantifiable, approach). However, the greatest danger, in our view, is that of ambiguity. A clear choice of one of the two approaches described here, on the other hand, results in a manageable split, with two different grading scales. It would be interesting to measure the differences between results obtained using the "everything" and the "outline" instructions – differences which may or may not be significant.

3.7. Tests and context(s)

Finally, in this section, we shall examine an issue which has only been touched on in passing so far. Contrary to what the authors of the tests described above would have us believe, there is no such thing as an acontextual language, and hence no means of testing a form of idealized, "pure" memory. For starters, the testing situation is far from neutral as a language context; outside of this context, it is very rare for a person to know that they are being observed for diagnostic purposes. Next, test themselves impose semantic constraints, justifiable or otherwise; as Rastier states (2001, p. 195), "every context plays a role in the actualization of semantic features. The name of a category is a context, as are the other examples in that category".

Thus, by asking a subject to list members of the "furniture" category and by giving the example "chair", a neuroscientist provides the patient with a clear framework, activating a sorting process for *furniture* versus *non-furniture* which is broadly irrelevant in a real-world communication situation. We rarely need to separate tables, chairs, etc. from all other things. On the other hand, *stool* might be opposed to *chair* by the semantic feature/comfortable/, for example, in the sentence "I have backache, I'll sit in the chair for lunch today". This example, featuring a *differential* seme, highlights the inevitable problem at the heart of all of the cognitive sciences. Meaning is always seen to be *referential* in nature, and this explains the predominance of categorization within the discipline; a signifier must correspond to a referent, and since referents are arranged around prototypes, the same is considered to be true of words.

Modern semantics[21], on the other hand, has shown that meaning is a question of *differences*, which constantly need to be (re)constructed and are determined by moving from the broadest context – social practice – to the narrowest context – the syntactic group in the sentence. In other terms, any categorization separate from a set of definitions (a context) is illusory. Note that the "connectionist" hypothesis has had a certain influence even within neuroscience. This position, notably adopted by Seidenberg and McClelland (1989) and Gaskell and Marslen-Wilson (1997), is summarized in the following manner by Spinelli and Ferrand (2005, p. 27):

> "In other words, according to this understanding of language, the memory does not store lexical representations; instead, orthographical, phonological and semantic codes are calculated on demand each time a word is read or heard".

McClelland (2000) himself, in a chapter on the various visions of memory, provides a brief presentation of the main principles of connectionist models of memory, sometimes known as Parallel Distributed Processing (PDP). This model centers on the idea that memory should be seen as interactions between perpetually changing neurons which react to given commands. In practice, the latest recent representation of an event may be more or less recent, but we cannot say that it is stored as such. The memory of this event is more like an association of information elements exchanged between neurons. When the memory of an event is recalled, it is, in fact, this association of exchanges which is recalled, or, more precisely, reconstructed, as the memory is located in connections between neurons. The most iconoclastic aspect here is the idea that individual memories are not stored in a single location; everything comes down to remodeling and reconstruction (McClelland 2000, p. 593):

> "Under some circumstances – for example, when the constructive process takes place in response to a recall cue – the cue may result in the construction of a pattern of activation that can be viewed as an attempted reconstruction of the pattern that represented the previously experienced event. Such a

21 In spite of chronic underfunding, which the cognitivists, hanging on the coat-tails of Chomskyism, have not had to face.

> reconstructed representation corresponds to a recollection. The patterns themselves are not stored, and hence are not really 'retrieved': recall amounts not to retrieval but to reconstruction".

Although the principle of perpetual reconstruction of events is novel, the examples given are not always convincing. Commenting on the classic one-way, top-to-bottom concept tree idea, which broadly follows the division of species with a few added properties, Rumelhart and Todd (1993, p. 15) proposed an alternative, network-type model, connecting input concepts on the left with output properties on the right. These units are linked through a set of connections (e.g. "is a, is, can, has"), and through the strangely named "general hidden units" area, shown on the right of the concept zone (in the second series of circles) in Figure 3.7, which reads from left to right. The input is "robin", which activates the "can", "grow", "move" and "fly" units. This figure is a simplified version of the authors' original diagram, in which all units are connected by multiple lines, making it hard to read; we have simply retained those connections required to illustrate our point.

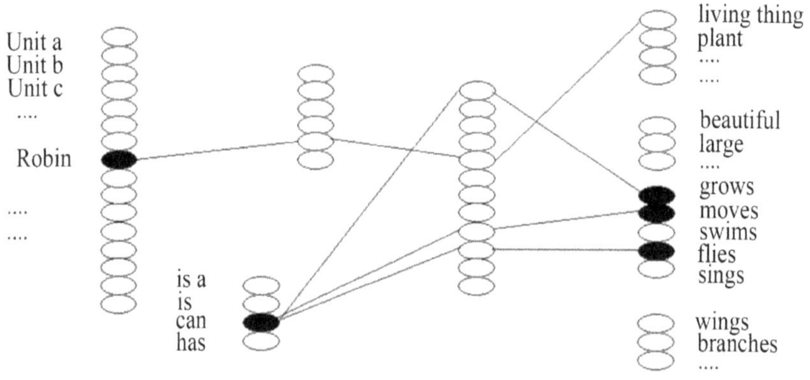

Figure 3.7. *A connectionist semantic network (adapted from Rumelhart and Todd (1993))*

Figures of this type clearly resemble finite state automata, and this is part of the reason we find them disappointing. Although it does not pretend to be anything more than a remodeling of a concept tree, the network is closed in on itself, greatly limiting our options. In our opinion, there is a serious deviation between the stated ambitions of connectionist theory and representations of this type.

In any case, it is now widely accepted that what we call "memory" is the crystallization of traces of neural connections. Many experiments, using fictional or formal neuron networks, have shown that the image of items of information and/or memories as being like a dried flower or an old postcard in the back of a drawer is not relevant; the "storage spaces" themselves are inscribed in the connections.

The idea of a perpetual reconstruction, not of events in memory, but of units of language corresponds precisely to what is predicted by textual semantics, in which each occurrence is unique, since it is always situated in a new context. This does not eliminate past occurrences, but every new occurrence of a word implies an interpretative movement of contextual actualization. Reconstruction and reinterpretation occurs with every sentence. In view of what we have seen so far, it is no coincidence that cognitive semantics never deals with puns or other word games that can only be understood via interpretative paths or operations. How can we understand "a man's a man, for all that" if we consider that the word "man" has the same exact sense each time? It is doubtful whether even the most recent connexionnist models have the capacity to provide effective responses without leaving room for context. Interpretative semantics, which uses the hermeneutic principle of determination of the local by the global, certainly does not lend itself to models like that proposed by Rumelhart and Todd; that said, it is also immune to the consequences of this omission.

We thus agree with Spinelli and Ferrand (2005, p. 91) that "the semantic context must be sufficiently limiting to have an effect on the early stages of lexical activation". However, more generally, the minimalist idea of "context" used in psycholinguistics is regrettable; most of the time, context is limited to variations on the syntagmatic level alone: for example, in French, "le/la [N]" where [N] is a word such as *vase*, which has different meanings in the masculine and feminine (*le vase*: vase, e.g. for flowers; *la vase*, sludge). In this case, the two meanings do not habitually occur together. Another example is "[adj.] plane" (*large* vs. *genuine*, to study reaction time with a true/false /t/ link); in reality, however, interpretative operations are habitually carried out on at least a phrasal basis.

3.8. Absence of a cultural dimension in cognitive testing

Similarly, we have also touched on the fact that existing cognitive tests do not include a cultural dimension. While there is no clearly stated opposition to a conception of language as a social product relative to a given culture, the new cognitive disciplines tacitly ignore this aspect, forgoing the fruits of a century of reflection and debate on the question. When cultural differences are discussed, it is only to minimize their significance. The insufficiency of this approach is clearly visible in two examples, taken from our experience of observing consultations.

3.8.1. *First-generation North African patients in France*

Neuropsychologists working at a hospital in R. (France), a densely populated area with a high immigrant population, have noted the importance of cultural disparities. To illustrate the potential impact of this cultural variation on testing practices, we shall consider the example of first-generation North African immigrants, who emigrated from their countries of origin (generally Algeria or Morocco) as adults. Most of these individuals continue to speak Arabic at home, a fact which limits vocabulary assimilation. The problem here, which has also been noted by speech therapists in different testing contexts, is that patients do not actually know the word they are supposed to find when presented with an image, or have only encountered it passively. Naming tests are therefore biased from the outset. Similarly, subtle variations in cultural codes of representation mean that some individuals may have more difficulty than others in recognizing the objects drawn. Practitioners have thus developed the habit of neutralizing certain points which, in their experience, are problematic. By their own admission, the solution is not entirely satisfactory, but it avoids an artificial reduction in patients' scores. It also provides something of a response to what Ollivet and Debrut, in a short article, call the "double exclusion" (Ollivet and Debrut 2007, p. 139), based on patients' status as members of an ethnically characterized group (Afro-Caribbean, Arab, etc.) and to the disease itself.

3.8.2. *Erasure of Czech/Slovak cultural disparities*[22]

In Slovakia[23], doctors and psychologists refer to different standards when assessing Hungarian and Romany patients. This situation is particularly strange as professionals in this country do not have access to tests specifically developed for Slovak speakers.

In practice, all of the tests that psychologists or doctors use to diagnose neuropsychological diseases are of Czech origin. Psychologists base this approach on the assumption that the language and mentality of Czech and Slovak patients are similar. However, there are two problems with this observation. Firstly, even if we consider that Czech and Slovak "mentalities" are similar, can the same be said of the language, which forms an inextricable part of testing?

There are identifiable differences between Czech and Slovak on every level. Firstly, certain letters exist in one language but not in the other: (ř, ou, ě, ů in Czech vs. ä, ia, ie, iu, ô, etc. in Slovak). In terms of morphology, different cases are used: for example, "to manage to…" is constructed using the genitive in Czech, "dosáhnout něčeho", while the accusative is used in Slovak, "dosiahnuť niečo". Certain verbal locutions are constructed differently on a syntactic level: take the expression "I miss you". In Slovak, "chýbaš mi", the subject of the phrase is the second person, as in the French *tu me manques*, but this "person" is not directly mentioned. In Czech, on the other hand, the construction "postrádam tě" corresponds to the English "I miss you", where the subject is the first person, and the second person is present as the object of the phrase.

Secondly, while the two languages are mutually comprehensible, there are often distinct words ("cushion" is "polštář" in Czech and "vankúš" in Slovak, "stork" is "čáp" in Czech and "bocian" in Slovak, etc.)[24]. All things

22 This section closely follows the outline of an article which we wrote in collaboration with Lišková (2018). The remarks made here draw heavily on the work of our co-author.

23 Our discussion is based on tests used to diagnose memory impairments (and the results of these tests) generously supplied by the Philippe Pinel psychiatric hospital in Pezinok (Slovakia).

24 Similarly, one wonders whether the linguistic manifestations of Alzheimer-type dementia are similar in languages which are more typologically distant. To cite just one example, do speakers of the Bantou languages lose the use of classifiers, and if so, to what does this correspond in languages where this category does not exist?

considered, the adaptation of tests to a culture and language group should not be controversial, but standard. In reality, however, this aspect is often overlooked. As we know from a hospital context, many of the tests used to diagnose neuropsychological diseases were originally developed by American researchers who prohibited psychologists or doctors in other countries from adapting them to their needs; tests could be translated as-is, word for word, or else not at all. This left practitioners with no choice but to work with the American tests, often without access to the official version (prohibitively expensive), which they administered in Slovak. Given this obvious bias, many practitioners, including psychologists and speech therapists, have made clandestine modifications to parts of the tests. The situation is gradually improving, but the case of Slovakia – where Czech tests are still used, on the assumption that, as the two countries were formerly united, population mentalities are essentially identical – shows that there is still a long way to go.

3.8.3. *Connecting cognition and culture: origins and perspectives*

As we have seen, there has been some improvement with regard to the integration of cultural disparities. For example, *Gremoire 1* and *2* both specify, in many cases, whether a translation into another language is available. As one might expect, however, all of the languages in question are Indo-European, and it is clear that standardized psychometric tests assume a "Western" cultural context. We would especially draw attention to the fact that these shortcomings can be attributed to the *universalist* approach inherent in the cognitive sciences. Following Chomsky, and the idea of an innate and universal language learning program, the first cognitivists – such as Berlin and Kay in the 1960s, then Rosch – were forced to counteract the potentially destructive effects of linguistic *relativism* on their field[25]. Writing

25 "The world is presented in a kaleidoscopic flux of impressions which has to be organized by our minfs – and this means largely by the linguistic systems in our minds. We cut nature up, organize it into concepts, and ascribe significances as we do, largely because we are parties to an agreement to organize it in this way. [...] We are thus introduced to a new principle of relativity, which holds that all observers are not led by the same physical evidence to the same picture of the universe, unless their linguistic backgrounds are similar, or can in some way be calibrated" (Whorf 1940, pp. 213–214).

in 1921, Sapir stated that "morals are the result of what a society thinks, language is the particular manifestation of thought" (Sapir 1970, p. 262); little did he know that a whole generation of researchers would attempt to limit the scope of this position by defining areas of exclusion, such as the so-called *natural categories* and, more generally, anything related to universals *naturally* resulting from human functions (Rosch 1977).

As we saw from our presentation, it is this very conception that tacitly underpins the design and construction of psychometric tests. As such, the non-expressed and undiscussed character of the assumptions makes it highly unlikely that psycholinguistics and neuropsychology will change from the "inside out" and produce tests emphasizing a differential logic of meaning, accepting categories (such as *furniture*, *trees*, etc.) in a communicative situation. Evidently, it would be harder to set up and administer these tests, and they might raise new and unforeseen problems, but they would give us the ability to measure a whole host of unexplored parameters.

3.9. Summary: rewriting tests

The marginalization of textual and cultural dimensions in testing is the result of a generalized conformation to enshrined postulates that are difficult to update, as well as, doubtless, to the practical constraints involved. Neuroscientists need to limit the variability of possible interpretations of the instructions, as well as the variability of possible, acceptable answers, to the greatest possible extent. However, by ignoring social practices and the inherent textual nature of language – evident in the way we define the meaning of words to a learner or a child, intuitively reconstructing contexts of use – we also reduce our capacity to measure patients' ability to retrieve information using practical and contextual clues. What neuropsychologists refer to as "delayed recall" is, in fact, nothing more than the same question asked 5 or 15 minutes after the first iteration; at best, a category name may be provided to assist patients in remembering. In this way, tests occur in an idealized, acontextual fantasy world.

Above and beyond the fact that a refusal to take account of recall cues constitutes a severe infringement on the hermeneutic principle, operating in both production and reception, we should note that these cues could be extremely helpful in differentiating pathologies. We know that patients with Alzheimer's disease do not seem to benefit from the use of recall cues, unlike patients with vascular diseases (see cases below). Two patients attending a consultation for apparently comparable memory complaints may respond in opposite ways in "assisted" recall situations, depending on the pathology profile.

CASE: PATIENT 1.– 79-year-old woman, left school before the age of 15, then worked in manual employment until retirement. Has family problems but does not express suffering from them, has never been married. Referred to the neuropsychologist by her usual doctor. Brought in by an assistant working for a personal support association. Does not admit to her own impairments and denies forgetfulness ("I know, I just can't say it") along with difficulties in both spatial and temporal orientation.

A full range of tests was carried out, during which the patient made multiple errors (in recognizing and reproducing drawings), notably including the following example: asked to name a drink, she mentioned the word *cognac*, but proved unable – despite having chosen the word herself, and having stated she does not like it – to recall the word on first, second or third attempts, even when the neuropsychologist provided the "drink" category cue on each occasion. The MRI revealed vascular problems, which we expect to cause encoding problems but which do not necessarily result in such a rapid loss of well-encoded information. Only time will tell whether this is the beginning of Alzheimer's disease, or if the woman's symptoms result from a different pathology.

CASE: PATIENT 2.– 74-year-old woman, guitar teacher. Graduated technical high school. Friends have noticed an increasing number of mistakes over the past year. The patient accepts this, and mentions the fact that she has always had difficulty concentrating.

She, too, completed an in-depth neuropsychological assessment and appeared to have difficulty remembering words: on occasion, she made remarks such as "you don't see one of those every day" when faced with animals or objects which she was unable to name. Generally speaking, she worked slowly and clearly had difficulty completing the tests without assistance. Unlike patient 1, however, she was systematically able to recall words with ease once a category cue was given. Following the examination, she said that she was extremely anxious and suffered from sleep apnea.

In our opinion, the use of more, better-designed recall cues might assist with differential diagnosis. The value of this approach is confirmed by data found in the literature: Dubois *et al.* (1997, p. 154) state, in an article concerning exclusively episodic memory, that while "performance in free recall is significantly impaired in Alzheimer's disease", "performance in indexed recall varies very significantly between patient subgroups". Since the authors of the article neglected to comment on the two tables provided on page 155, readers are left to reach their own conclusions; our own remarks are given below, although we have not included the raw data. The authors' idea was to cross-reference different "characteristics" of Alzheimer's patients: their age, their MMS exam score, their score on a test designed to evaluate depression and finally socio-cultural level (always hard to determine with certainty). In this way, they divided the patients into four groups of 30–40 patients. These groups ranged from a first cohort of patients with an MMS score of over 26, aged about 70, at the highest socio-cultural level and with a low depression score, to a group with very low MMS scores (under 18) where all other variables were comparable to the values for the first group, with the exception of age (average: 74, common to groups 2–4).

Although the authors present several criteria, their grid is essentially based on MMS scores (shown at the top of the column), and it is clear why this is the case. There is a visible decrease in all results as the MMS descends, for immediate indexed recall, free recall or "total" recall, where cues are multiplied until the target word is recalled. The most we can say is that when the results for the third group reach a very low level, the results for the fourth group do not demonstrate a comparable decline. Unsurprisingly, patients in the earliest stages of Alzheimer's disease are the most sensitive to recall cues. At the outset, these patients appear to behave normally; they sometimes forget things, but will remember as soon as a recall stimulus is presented. This is an aspect which family and friends often notice. In all cases, sensitivity to recall cues appears to be a strong indicator for Alzheimer's disease.

Clearly, altering all tests in this way, using a variety of real contexts, would pose significant difficulties, for example, in terms of test standardization and in establishing a scale of values. It would also take longer to carry out each assessment, with a knock-on effect on costs. It is equally evident that, as linguists, we lack the skills to match diseases with types/frequency of forgetfulness. However, this pathway does seem promising. For example, patients could be assessed via a narrative context, using a text such as a short story to see if they can recover words which they could not produce in naming tests, either immediately or with delayed recall. If patients are seen to have "lost" certain words or even social practices in tests, we could assess whether they are capable of "finding" them again in a clearly structured conversation situation, such as a role-play scene. It would be entirely possible to mime a cooking scene with patients in which these linguistic and cultural skills play a part. Although this may be hard to implement in practice, we have seen that limiting recall cues to object categories alone is not sufficient.

More simply, smell or taste stimuli could also be used, allowing us to see whether all of the senses are affected at the same time and in the same way. The fact that this angle of approach is not used in tests may be seen as yet another symptom of the overwhelming dominance of the visual aspect in Western society. Observations made in this way could be used to see whether enriched encoding, or recall in a more "natural" context, help certain patients more than others; this is a stated aim of some of the most recent tests. Allain *et al.* (1997, p. 192), writing over 20 years ago, felt that a more "natural" style of testing was beginning to emerge, but few changes have been seen in practice, at least as far as language is concerned:

> "Many psychometric tests enable a fine and repeated exploration of the different components of human memories, alongside other cognitive functions (executive functions, speed of understanding and processing information, etc.). Several of these tests have a proven ecological value, i.e. significance in daily life".

Natural language is the means of communication used by the majority of tests, and shared language is a necessary condition of assessment. The speed of understanding, to cite just one example, cannot be evaluated without the use of language, making it imperative that we develop a clear and accurate understanding of the language itself.

Thus, in response to Laurent and Thomas-Antérion's question (1997, p. 104) concerning the best tests to use in detecting early-stage Alzheimer's, we might suggest, from a linguistic standpoint, making adjustments to all of the available tests. While the systematic use of long interviews, as practiced by Alois Alzheimer himself, seems impossible to reconcile with the increasingly large number of patients, we feel that a framework featuring a greater number and variety of contextual clues, designed to be as fun and informal as possible, would help to eliminate those errors resulting from the nature of tests themselves and from patient stress, making it easier to spot the effects of the disease itself. We do not expect our suggestions to be universally accepted in the neuropsychology world – far from it – but we feel that these points could form the basis for discussion and for an exploration of new pathways.

4

Analyzing the Speech of Alzheimer's Patients: Methods and Perspectives

More and more linguists have been working on establishing the foundations of what might be called "corpus linguistics in clinical context". Since the 2000s, *corpus linguistics*, in the English-speaking world, and *textometry*, in French-speaking circles, have revolutionized textual analysis: by analyzing texts on a quantitative basis, textometry has redrawn the map and reinforced the hermeneutic principle of determining the local by the global. To take just one example, we now know for certain that the genre variable takes precedence over the author variable; two texts of the same genre are likely to have more linguistic properties in common than two texts by the same author. These gains have only been made possible by the existence of large corpora, often written (literary or journalistic); at present, nothing of the sort exists for patient speech. Nevertheless, some isolated attempts have been made, following the tradition initiated by early psychiatrists such as Alois Alzheimer, who spent long periods of time with his patients, recording his observations in shorthand.

4.1. An eclectic range of approaches

The work that is currently ongoing in this area seems, to us, encouraging: despite limitations in the size and variety of the available corpora, they demonstrate a clear willingness to analyze texts in a multi-factorial and quantitative manner. One notable example is a study by Cantos Gómez (2010), entitled "Analyzing linguistic decline in early stage Alzheimer's disease: a corpus-based approach". This study forms part of a book exploring the state of the art of corpus linguistics and highlighting recent developments, including medical applications. Like all linguists working on neurodegenerative diseases, Cantos Gómez starts from the notion that Alzheimer's disease is characterized by memory loss, and that it must be possible to find quantitative traces of this loss in patient discourse, as long as we have the required idiolectal indicators to characterize this evolution. The author focused his study (Cantos Gómez 2010, p. 168) on the "spontaneous speech[1]" of Harold Wilson, a former British Prime Minister who suffered from Alzheimer's disease. To do this, he drew on the work of Garrard (2009), who claimed that, by the end of his political career (1975–1976), the content and themes of Harold Wilson's speeches had lost their distinctiveness compared with those of other members of parliament – a distinctiveness which was particularly notable in the years 1965–1966. Garrard used two variables: the 30 most frequent words and 30 *content words* or lexemes. According to Cantos Gómez, however, this study has shortcomings, particularly in terms of the precision of the linguistic variables. In our view, it would have been better to measure the semantic networks developed by the 30 lexemes; Cantos Gómez did not take this approach either. However, the study is still valuable; the corpus contains around 30,000 words for each of the analyzed periods, and the major findings are described below.

First, with regard to the prosodic segmentation of discourse, Cantos Gómez notes a very clear decrease in commas (by around half), semicolons (four to five times fewer) and question marks (six times fewer). Wilson's use of deictics (2010, p. 177) is also surprising. Indeed, singular demonstratives were under-used by Wilson in 1976, whereas place adverbs were over-represented in his speech. Oddly, he also used greater numbers of conjunctions (both subordinations and coordinations).

However, at first glance, the author seems to contradict all the observations made about the extent of vocabulary loss in Alzheimer's patients, detecting no decline before 1976, when Wilson resigned, there was no statistically significant loss of skills with respect to either grammatical or lexical words. At most, we might detect an increase in the repetition of

1 Excluding declarations, which he always read from a paper and which were often written by others.

common nouns. However, the devil is in the details, and the most interesting part of this study lies in the following remarks (Wilson 2010, p. 181): over time, Wilson's use of rare words becomes increasingly limited, resulting in an increase in *hapax legomena*; on the contrary, common words, particularly hypernyms, feature to a greater extent in Wilson's speech over the period 1975–1976, and this is consistent with what has been recorded as a common early symptom of Alzheimer's disease. Cantos Gómez notes that Wilson was aware of his emerging difficulties and must have had a significant cognitive reserve allowing him to use a variety of different linguistic pathways to express his ideas. When he no longer felt able to do this, he resigned. As we can see, this survey – although limited in scope, due to the size of the corpus and its idiolectal character – indicates a number of strictly linguistic and statistically justified pathways for further investigation.

Cantos Gómez's study is complemented by Devevey and Balavoine's observations on *neoforms* – words that may be defined as unattested lexies which are morphologically correct and also occur in a suitable syntactic position. These authors mention that, in patients with late-stage neurodegenerative diseases (including Alzheimer's), "the lexicon becomes poorer over time (as confirmed by data in the literature) and syntax tends to become simpler, but more importantly, at the same time, there is a significant increase in the number of neoforms per utterance" (Devevey and Balavoine 2010, p. 337). Nominal forms (Devevey cites "gnoquets" in "c'est des gnoquets") and verbal forms ("caper" in "j'veux pas caper") feature most heavily, and are provided with correct derivational and inflectional morphemes, respectively. This first fact is interesting in itself, since it indicates that grammar remains intact. Furthermore, neoforms are more or less interpretable, leading the authors to suggest that these may represent "the beginnings of idiolectal semantics" (Devevey and Balavoine 2010, p. 340). This is not surprising, as the meaning of neoforms can often be discerned from either the immediate or wider context. This study highlights an aspect that is often overlooked: as the shared lexicon is increasingly impaired, we see the emergence, then a clear increase, in the proportion of idiolectal neoforms in patient speech. Note too, in passing, that the interpretability of these neoforms runs contrary to the notion of an acontextual lexicon.

On the basis of this observation, note that we strongly oppose the old and well-established notion that language is, essentially, composed of a dictionary and a grammar, sometimes with a few key texts for good measure.

As Rastier (2013, p. 151) notes, “languages have all too often been reduced to dictionaries and grammars, or even syntaxes”. The groundwork for the separation between grammar (based on logic) and interpretation (rhetoric) was laid by Aristotle, and this division that is hardly contested even today in academic circles. European missionaries to far-off lands began their linguistic endeavors by establishing a bilingual glossary – for example, when faced with a seemingly impenetrable Amerindian language. The aim in this case was rather specific – to translate the New Testament into the language of native populations as quickly as possible. However, it still indicates that the dictionary was seen as the basic, and perhaps most important, document. Diagnostic tests are influenced by this notion that grammar and lexicon are two separate components: practitioners attempt to determine which of the components is best preserved in patients, meaning that one aspect of the analysis grid is taken for granted. The idea of combining the two, in the guise of interpretation – a dynamic process in which grammar and lexicon interact – is only rarely expressed. In short, neuropsychological tests feature a componential representation of the lexicon, as a module to which in Isaac Asimov’s short story *Someday* (1956). Niccolo, who has an older version of the “bard” – a robot with the capacity to invent stories at will – complains that his model only produces tales of kings and princesses. His friend Paul offers to make a few modifications:

> “Paul, who wasn’t listening to the Bard, said, ‘It’s *easy*. The Bard has memory-cylinders all fixed up for plot-lines and climaxes and things. We don’t have to worry about that. It’s just vocabulary we got to fix so it’ll know about computers and automation and electronics and real things about today. Then it can tell interesting stories, you know, instead of about princesses and things’. […]
>
> Paul pulled out a thin, flexible metal strip, powdered with dots. ‘That’s the Bard’s memory cylinder. I’ll bet its capacity for stories is under a trillion’.
>
> ‘What are you going to do, Paul?’ quavered Niccolo.
>
> ‘I’ll give it vocabulary’.
>
> ‘How?’
>
> ‘Easy. I’ve got a book here. Mr. Daugherty gave it to me at school’.

> Paul pulled the book out of his pocket and pried at it till he had its plastic jacket off. He unreeled the tape a bit, ran it through the vocalizer, which he turned down to a whisper, then placed it within the Bard's vitals [...].
>
> 'What'll that do?'
>
> 'The book will talk and the Bard will put it all on its memory tape'".

This concept seems attractive, but it belongs firmly in the realm of science fiction: without global data on the use of lexical units, or with access to just one single book, the Bard would find it difficult to use words correctly – even though it was able to segment text beforehand by isolating, for example, the idiomatic expressions that constitute at least one tenth of any European lexicon. Asimov, who was both a pluridisciplinary scientist and an inventive storyteller, thus retained a remarkably traditional vision of language.

To the best of our knowledge, only one study has gone further in this regard. In *When language breaks down – Analyzing discourse in clinical contexts*, Asp and de Villiers (2010) start with the observation that very little quantitative data is available concerning speech in a clinical setting, despite the fact that "[...] people interested in neurological disorders and diseases should have access to very detailed accounts of the discourse patterns (and other behaviors) of the population they study and treat" (Asp and de Villiers 2010, p. 2). Their work provides a rich supply of tables and speech extracts, most of which are taken from interviews with autistic patients, although some feature Alzheimer's sufferers, making them all the more valuable for our purposes.

Asp and de Villiers (2010, p. 4) begin by examining how linguists can characterize discourse which deviates from the norm, considering the following text in which a patient talks about her mother:

> "Ah the same thing with the – what was it just discussed? – the stove. I can remember my Mom, she was as smart as a ticket. She was ninety-six years old when she died and she would, you know, she has everything under control. But I knew that I couldn't do that I couldn't go and I don't know. Now I'm lost again".

This 76-year-old patient forms grammatical sentences and, contrary to research findings on the loss of abstract thinking, still uses phrases such as "as smart as a ticket". However, the fact that she has difficulty remembering the theme of the question (the "stove") and ends by confessing that she is confused betrays a loss of focus which is characteristic of Alzheimer's disease. Asp and de Villiers aimed to find out whether there are other less visible, but equally relevant, symptoms, taking a theoretically eclectic approach which combined conversational analysis (speech analysis), generative grammar, functional linguistics and pragmatics. In our view, this eclectic strategy is not the strong point of the work, and it is hard to see how all of these theoretical currents can be made to fit together. From another angle, however, the authors never attempt to synthesize all of these views, and it soon becomes apparent that their aim is to develop a palette of complementary analytical tools. Once we accept this approach, there are several relevant points to draw from Asp and de Villiers's work.

The list of objective criteria which could be used to identify deviations from normal speech begins with different types of *confabulations*, which are described as follows:

> "Story elements are judged to be confabulated when processes, participants or circumstances (or entire predications) which are intended to represent story elements are reported but are not present, nor synonymous, with the control narrative".

As we saw in relation to the Lion's Tale, there are no criteria for identifying this impairment. However, Asp and de Villiers use extracts to illustrate a number of distinctions which are commonly accepted, including at least three types of confabulations which are particularly prevalent among Alzheimer's patients. The first consists of wrongly identifying a source of information: for example, the patient may believe that information has been transmitted by an individual, when in fact they remember it from a radio program or other source. The second type is well known to any neuropsychologists working in the field: Alzheimer's patients have a propensity to invent purely fictional elements concerning their personal past: this is the second type. The last type concerns the wholesale replacement of data (actors or circumstances) with other data, for example misunderstanding the time of day when a story takes place. In short, a distinction can be made between "pure" confabulation, in the sense of invention (2) and confabulation resulting from subconscious confusion, whether about the source of information (1) or its content (3). The nature of confabulations may be less important than their frequency, as the authors suggest, but this is not certain: this hypothesis could only be verified with access to a quantity of data which is not yet available, due to the lack of a sufficiently substantiated corpus.

Asp and de Villiers also confirm Duong's findings concerning the over-representation of *tags*, particularly question tags, in the discourse of Alzheimer's patients: these "adverbial" phrases, such as "isn't it" or "do you", whether within or at the end of a sentence, may indicate hesitation, or show that the patient is seeking approval from their conversation partner. Using a corpus of around 100 interviews with caregivers, researchers have demonstrated that Alzheimer's patients use question tags with significantly higher frequency; however, the range of tags used by these individuals is relatively small.

Patients with ATD have a strong tendency to doubt the reliability of their own memories, particularly episodic memory, and hence make frequent use of question tag such as "do/did I?" (e.g. "I don't have a dog, *do I?*"). Over 90% of occurrences of this type in the corpus are uttered by patients, with the remainder coming from caregivers; furthermore, caregivers tend to use these tags mostly when talking to patients about what they did together in the preceding hours or days ("did we?"). The use of "autopersonal" question tags is a strong speech marker in English-speaking Alzheimer's patients. French patients do not appear to use "*n'est-ce pas*" to the same extent, and we consider that these observations are language-specific. This is not to imply, however, that the findings discussed above cannot be applied to French, as long as the paradigm is extended to all elements uttered with a rising intonation. To obtain comparable results, "*non?*", "*tu ne crois pas?*", "*hein?*", etc. would also need to be taken into account. The problem would then lie in isolating those elements of the paradigm which are "autopersonal" in nature, something which is easy in the case of "did I/did you". Following this same logic of self-questioning, Alzheimer's sufferers use an abnormally high number of modal verbs and conditional verbal structures, verbs associated with thought and doubt, such as "believe" and "suppose". In the following conversation extract[2], the patient seems to experience doubt when trying to

2 Author's note: as in the previous chapters, where text is to be analyzed, the original French text is shown alongside the English translation.

explain the reasons he got lost. His speech turns feature a high degree of tension between /certainties/ ("*j'connais bien*", "*tous*") and /lack of understanding/ ("*j'sais pas*", "*j'crois*", "*je crois*"):

(1) NP: //Comment ça s'passe à la maison?//	(1) NP: How are things at home?
(2) C: ((breath)) //Ben//c'est d'venu assez difficile//Il s'est perdu quand même un peu quelques fois dans l'quartier//Il est revenu plus tard que prévu//	(2) C: Uh, it's got quite difficult. He's got a little lost a few times in the neighborhood, he came back later than expected.
(3) NP: //D'accord//M.L., qu'est-ce qui se passe quand vous sortez?//Est-ce que vous avez des difficultés par exemple à utiliser les moyens de transport//J'sais que vous avez l'habitude normalement d'utiliser le bus//	(3) NP: Okay M.L., what happens when you go out? Do you have trouble using transport, for example? I know you normally take the bus.
(4) P: //Ah ben j'connais bien tous les changements//et j'sais pas ce qui s'est passé//euh//le tram//il a #j'crois bien qu'ils ont changé des stations//Ouais, ouais//##enfin, je crois, euh//[3]	(4) P: Uh well I know all the changes and I don't know what happened, uh, the tram, it … I think they changed the stations, yeah, yeah. Well, I think so, uh.

This extract features a number of utterances, such as "*enfin*", which act as autotags, but which might not have been identified as such at first glance.

If we look closely, all these linguistic properties are indicative of a loss of control, first in memory and then in language. It seems likely that the

3 NP = neuropsychologist; C = caregiver (in this case the patient's wife); P = patient. *La maladie d'Alzheimer*. Documentary extract, University of Geneva [Online]. Available at https://www.youtube.com/watch?v=oJYx7c-52D0 (published September 4, 2012), min. 8:24–9:06. The original, annotated text is shown on the left, and a non-annotated English translation on the right.

opposite trend is observed in the discourse of non-native speakers who are learning a language. Initially, tags, irrelevant insertions and redundant presentational additions are used, decreasing in frequency as mastery of the new language increases. This observation opens up interesting perspectives for a research project, identifying substitution strategies which speakers use when they encounter difficulties, whether or not these are linked to a pathology.

Morphosyntactic complexity and variety are also relevant in assessing patient speech. The use of subordinate clauses (notably relative and circumstantial) is costly in terms of working memory, in that they often require speakers to create co-reference links. The same cannot be said of passive structures, which simply imply an inversion of the subject and object. The rarity of relative subordinate clauses, correlated with other factors, may be significant. There is a lack of data with regard to cleft structures ("it was X who ..."), pseudo-cleft structures ("the one who ..., that was") and extrapositions on the left used to thematize a construction ("This, I don't want it"). However, it seems probable that all syntactically complex structures implying a chain of reference will be problematic. Caregivers and healthcare personnel could be made aware of these elements to improve communications with patients. For example, it could be wise to limit the number of subordinates and to maximize the use of "simple" sentences, in the propositional sense of the word, that is, sentences with only one clause.

Lexical density could also be cited as an indication that discourse deviates from the norm. While some authors, such as Ure, calculate density by multiplying the number of lexical words by 100 and then dividing the resulting number by the total number of words, Halliday (1990) chooses to divide the number of lexical words by the number of clauses. Lexical density is known to be higher in written than in oral language (essentially because oral language features more occurrences of grammatical words, alongside to more numerous and compact propositions). However, this variable has yet to be tested in comparing the speech of an Alzheimer's patient with that of average speakers. Lexical density in Alzheimer's patients appears to be slightly lower than the value of 1.8 which Halliday cites as average in spontaneous adult speech, but not significantly so.

Element	Category	Features	Number	Value
Speaker turns	Independent clauses	(include elliptical responses to polar questions)	12/25	48% of all turns
			9/12	75% of patient's turns
	Minor clauses (lacking a subject or predicate)		3/12	25%
Speech functions	Statements	Full	6/9	66% of patients' independent clauses
	Polar questions	Elliptical (positive)	2/9	22%
		Episodic check	1/9	11%
Modalization	Statement check	Monitoring (*you know*)	1/9	11% of patient's statements and minor clauses
	Modal verbs	Hypothesis (*would*)	1/9	11%
	Mental cognition	Hypothesis (*guess, suppose*) *not know* (2) *think*	1/9	55%

Table 4.1. *Modalization features in the discourse of a person with Alzheimer's disease*

Furthermore – predictably – this appears to vary considerably depending on the theme of discussion, according to Asp and de Villiers (2010, p. 118). Alongside an assessment of the lexical richness of a patient's speech – and for this to be useful, we would need to know how it compares to their pre-pathology performance – repetitions and the use of highly general terms, targeting insufficiently defined referents or sets, constitute valuable cues. Patients with Alzheimer's disease have a tendency to doubt their own memory, and so often use more general concepts in the place of precise labels, for example: "neurologist" becomes "doctor", a precise object becomes "a sort of thing", a variety of verbs give way to "do", etc. The combination of these discursive elements result in what Asp and de Villiers

(2010, p. 132) termed "co-operative but 'empty' speech"[4]. Table 4.1 (Asp and de Villiers 2010, p. 131) shows discursive characteristics observed in an individual with early stage Alzheimer's disease, featuring the majority of the variables discussed above.

The over-representation of modal verbs is clearly visible in the final line of Table 4.1. We also see an almost equal distribution of speaker turns, although this tells us nothing concerning the nature of discussion: although the patient responds and, in the words of Asp and de Villiers, "cooperates", many clauses remain incomplete, elliptical, and lack a subject or verb. In a way, we might suggest that, given the high number of modalizations as signs of doubt, even the assertive nature of patient statements could be called into question.

Based on this state of affairs, it seems helpful to take a wide range of different approaches to "clinical context" discourse. The variety of theories applied to these specific corpora highlights a number of specific features, and while many appear logical and predictable, they lay the foundations for a differential overview of the discursive effects of various memory impairments in patients. It is also evident that sufficiently long recordings and well-annotated corpora are required in order to derive full benefit from this approach; precise and well-defined standards will also be required, such as those established by neurologists in the context of psychometric testing.

4.2. Patient–physician dialog and autobiographies: microsemantic analysis

In this section, we wish to show how our own discipline – interpretative semantics, notably microtextual applications – can contribute, in some way, to identifying linguistic markers of Alzheimer's disease. As we know, memory impairments are often the earliest symptoms of the disease, occurring well before apraxia and agnosia, for example. Clearly, it should be possible to observe imprecise referential expressions in discourse. The

4 "The pattern of co-operative but 'empty' speech in AD has been attributed variously to the breakdown of lexical–semantic systems, conceptual systems or to problems with executive function processes which limit access to these systems".

extract shown below, taken from an interview with a patient suffering from Alzheimer's disease (Asp and de Villiers 2010, pp. 129–131), includes a number of more or less fuzzy references on several levels: temporal ("the past year"), personal ("someone") or other ("something"). Unlike Asp and de Villiers, however (see Table 4.2), we do not think that this list is sufficient, particularly since the patient's aim is to describe a general and frequent situation in which "someone" told him that he had not done "something", which he thought he *had* done. Similarly, the list of thought verbs ("suppose", "guess", "think", "know") is not significant in its own right, but becomes important if these words are often found in conjunction with conditional modalities ("would") or negations ("don't"). When these cues are combined, "past year", for example, may be seen as an imprecise reference, as it follows "suppose". Similarly, "recently" and "not very long ago", which would not necessarily be significant in the discourse of a healthy speaker, become significant here when found in conjunction with "guess" and "would":

> (1) IV[5]: Mmhm.
>
> (2) So can I ask you to tell me a little bit about the time when your health changed.
>
> (3) When you started noticing you were having memory problems.
>
> (4) P: Well I I *guess* it *would* only be {(laughs)} *recently* and *not very long ago*.
>
> (5) IV: Mmhm
>
> (6) P: Um I suppose # in the past year
>
> (7) IV: Mmhm.
>
> (8) What sort of things started happening that made you aware?
>
> (9) P: Um # well um of course I *would think* that I had said or done *something*
>
> (10) and *someone would* tell me I hadn't and that sort of thing you know.

5 "IV" corresponds to "interviewer", "P" to "patient" and "CV" to "caregiver". The units mentioned above are shown in italics.

(11) Mmhm

(12) P: Nothing nothing very drastic but ah just.

(13) CG: The little everyday things wasn't it?

(14) P: Little little things.

(15) IV: Mmhm

(16) and what sort of took you off to the doctor the first time to check it out?

(17) P: Ah well just what other people were saying to me

(18) that I was getting forgetful and so on.

(19) IV: Mmhm.

(20) P: But ah of course ah just the ordinary doctor that doesn't # have much of a remedy for that. ((laughs))

(21) But ah I I went to my doctor anyway

(22) IV: And and did your doctor refer you to the clinic?

(23) P: Ah # # did he?

(24) CG: Yes

(25) P: Yeah

(26) CG: Yes Dr Smith did Dear

(27) P: Yes Dr Smith yeah

(28) CG: He wanted the Geriatric assessment test done.

(29) IV: Umhum.

(30) And how was that for you?

(31) Was that was that test ah

(32) P: I I I ah I *don't know*

(33) IV: Guess you went through it

(34) and maybe didn't

(35) P: *Didn't* even *know* it ((laughs))

Asp and de Villiers summarize their observations as follows. While the distribution of parts of speech is unremarkable, the opposition between defined/undefined references may be seen as scalar rather than polar.

Once again, we do not support the "atomizing" analysis approach; a category-based inventory and a decomposition of content on a unit-by-unit basis are insufficient, and better results can be obtained using combinatory semantics. Working from these extracts alone, we see that all interpretative cues need to be used in conjunction with others. This is not an isolated case; in the following extract (Asp and de Villiers 2010, p. 106), a case of delayed recall, thought verbs ("remember", "think", "know") are, again, combined with negations:

(1) P: // what was the second story//.

(2) IV: the story was about a road report

(3) P: // oh yes// *can't remember* it dear//

(4) IV: any little detail?

(5) P: //aah // had there been an accident//

(6) IV: umhum

(7) P: // yeah// a crash//

(8) IV: yeah

(9) P: // (I) *don't know* anything else// (accident) (15)// I *can't think*//

(10) IV: ok

(11) P: // oh you poor thing// do you have many like me//

(12) IV: oh, you're doing fine

(13) P: ((laughs))

(14) IV: you're doing your best. [Instructions]

(15) P: Oh, I see. That's good

Reference (n) = number of occurrences			**Taxonomic relations**			
			Nouns	**Verbs**	**Adjectives**	**Adverbs**
Personal pronouns	I (9) me (2) you (1) he (1)	Superordinate	sort thing people	do get (become) have go		ago
Definite reference	the past year that sort of thing the ordinary doctor my doctor Dr Smith (echo rep.)	Basic level		guess suppose think tell say (2) know (2)	long past little (2 – echo) forgetful ordinary	recently
Indefinite reference	something someone nothing (2) things (echo) other people a remedy	Subordinate	remedy Dr Smith (echo rep.)		?drastic?	

Table 4.2. *Lexical patterns in the discourse of a patient with Alzheimer's disease*

The [thought verb] x [negation], although predictable, is clearly a strong marker. Furthermore, while auto question tags, such as "did I"[6] imply that the speaker doubts their own words, the [doubt verb] x [imprecise spatial/temporal information] appears to be a characteristic feature of discourse in Alzheimer's patients. Repetition in the form of a second delayed recall exercise changes nothing, as one might expect. The following dialog features the same approximations and constructions (shown in italics) (Asp and de Villiers 2010, p. 106):

> (1) P: // The second story was the fella who got up at dawn to go on a trip// fishing// and the road report came on// and gave him a bad accident report// and that there'd been a huge pile up, and roads closed and traffic all rerouted// and that was the advisory to people not to go out// and he did// and it was a very disastrous day// *that's what I got out of it*//(laughter)// but *that's about* the gist of the whole story//
>
> (2) IV: (laughter) yeah
>
> (3) P: // I can't word it word for word//
>
> (4) IV: Ok.
>
> (5) P: // There's no way the brain'll ()//
>
> (6) IV: Not a problem. Just what you can remember.

Note the generalizing observations concerning the gist of the story. The patient notes that he cannot recall the story word for word, concluding with a clear expression of regret, indicating awareness of his memory impairment. This feeling of inability to access memories that should be there is a recurring feature. It causes a certain degree of distress in patients, both for themselves and for their families, friends and caregivers, as we see from the utterance "oh, you poor thing", addressed to a medical professional in the previous extract. Ploton referred to this trait in Alzheimer's patients as "narcissic vulnerability" or *atteinte narcissique*, notions connected to the relational isolation and anxiety resulting from the patient's situation (Ploton 2004, p. 67).

---

6 As we see from the following example (see Asp and de Villiers (2010, p. 159)): (5) Did I tell you about that IV? I was there for Christmas and into New Year yeah I guess I told you. (6) IV: this year? (7) P: yeah oh maybe I didn't. I don't know. When did you last come?

It is crucial to note the difference between "I think I don't know" or "I don't know" in conjunction with awareness markers and unconscious confabulations, or – which is perhaps worse – returning incessantly to the same ideas.

While we have only covered one textual genre so far, that of the interview, it is also interesting to consider patients' own stories, that is, autobiographical narratives. Clearly, this excludes stories told by family or friends, which are not intended to capture what patients themselves have to say, but recount the experience of living with Alzheimer's disease. It also excludes autobiographies that have been reworked by editors or that are too limited in their descriptive scope.

Furthermore, it is better to choose writings begun before the onset of symptoms, or at least before the diagnosis of the disease. Dubois, who began writing her diary *Mémoire, ma petite mémoire* on the advice of her speech therapist following diagnosis, is thus excluded; moreover, as this diary only covers a 2.5-year period, the temporal scope is too narrow to fully assess evolution. Finally, the fact that the author addresses her diary to her own "petite memoire" reduces the narrative and descriptive interest for our purposes.

Only one publication in French meets our criteria: Claude Couturier's diary, entitled *Puzzle, journal d'une Alzheimer*. The author began writing her diary well before being diagnosed with the disease, and the editor chose to leave the text unchanged. The writing is fluent, the descriptions are clear, and the reflections clearly express the psychological state of the diarist. The diary follows the classic format, but, as we might expect, the temporal reference points themselves are somewhat blurred. Fuzzy dates, such as "one day (?) in summer 1996", are not unusual. Evidently, the dialogism inherent in speaking of oneself is present, implying a form of pseudo-exteriority, but Couturier's style remains simple and natural: "je" remains "je", and there is no switching back and forth between "je" and "tu", a feature of Dubois' diary (relating to the fact that the author is writing *to* her memory). Couturier uses very little shorthand, and events are narrated using a normal proportion of verbs. In terms of tenses, the author switches between the *passé compose* – used for events in the recent past – and the present, often used to describe her emotional state; in short, there is nothing out of the ordinary in this regard.

In terms of meaning, there are several recurring semantic features. As Asp and de Villiers predicted (from their study of an interview with medical personnel), /confusion/ is clearly visible throughout the following extract. This essentially manifests in the form of verbs, all in the first person: *perdue* (lost) (2), *rendu compte* (realized), *ne trouvais plus* (couldn't find anymore), *cherché* (looked for), *ne vois pas* (can't see), *n'ai vu* (didn't see), *ne sais pas* (don't know), *ne reconnaissais pas* (didn't recognize), *perdre la tête* (lose my mind), *ai du mal* (had difficulty), *mélange* (mix up). There are also nouns (*vertige* (vertigo, dizziness)) and adjectives (*fatigue* (tired), *incapable* (unable)). Other predictable features include /imprecision/, starting with the question mark in the title "*Un jour (?) d'été 1996*"; we observe adverbial prepositional structures such as *à peu près* (about, around) (x 2), *exactement* (exactly) in combination with *incapable*, and finally, *à un moment donné* (at some point), which is less conclusive as it could feature in the speech of any healthy person. Moreover, there are obvious iterations: *encore une fois* (once again) (2) and *par moments* (at times) are temporal markers which clearly indicate repetition.

However, for Table 4.2 to be truly complete, we need to include the various textual markers of suffering, /privation/ in our annotations. The presence of large numbers of substantives, pronouns or noun constructions indicating incompleteness – *bribes* (bits), *pointillés* (dots) – or, worse, absence – *rien* (nothing), *trou noir* (black hole) – is more surprising when compared with the observations made by Asp and de Villiers[7]:

/imprecision/

(1) *Un jour* (?) d'été 1996

/itératif/ *[confusion/*

(2) *Encore une fois*, aujourd'hui, je *me suis perdue* dans le parking de

(3) Carrefour; j'étais avec Céline et déjà, la journée avait très mal

(4) commencé.

7 The English version of the following text will not be presented alongside due to complexities with the layout. Any reader requiring the English translation can find it in Appendix 2 at the end of this book.

/*privation*/

(5) Je ne me souviens que de *bribes*, comme si la journée avait eu lieu

/*privation*/

(6) en *pointillés …*

/*confusion*/

(7) Je me suis *rendu compte* que nous étions sur la route au bout de dix

(8) kilomètres (je conduisais pourtant) quand, avant un croisement, j'ai

(9) mis mon clignotant pour tourner à droite! Céline m'a dit:

(10) – Où vas-tu? on ne va pas vers le village, mais tout droit en ville!

(11) J'ai essayé de me rappeler notre départ de la maison, mais *rien*, le
/*privation*/

/*privation*/

(12) *trou noir*! Je ne me souviens pas avoir conduit pendant *à peu près*
/*imprécision*/

(13) dix minutes. Ça m'inquiète mais Céline me dit que j'ai conduit

(14) normalement jusque-là.

/*imprécision*/

(15) Puis, apparemment, les choses ont *à peu près* bien marché, mais je

/*confusion*/ /*imprécision*/

(16) suis *incapable* de dire ce qu'on a fait *exactement*: tout ce que je

/*imprécision*/ /*confusion*/

(17) sais, c'est qu'*à un moment donné*, j'*avais l'impression* d'avoir

/confusion/

(18) le *vertige* dans les rayons du supermarché et dans la galerie

(19) marchande où nous aimions nous balader toutes les deux en

/confusion/

(20) flânant tranquillement. J'ai dit à Céline que je *devais être fatiguée*

(21) et qu'il valait mieux rentrer à la maison …

/confusion/

(22) Mais en sortant, je *ne trouvais plus* les clés du camion dans mon

/confusion/

(23) sac; j'ai *cherché* dans toutes les poches de ma veste, rien.

(24) Céline m'a regardée en riant et m'a dit:

(25) – Oui, tu dois être fatiguée… tu as les clés à la main.

/confusion/

(26) – Bien sûr, mais je *ne vois pas* le camion; où est-ce que je l'ai

(27) garé?- Il est là, regarde, juste en face … à côté de l'abri à

(28) chariots … Qu'est-ce que tu as?

/confusion/

(29) Je *n'ai vu* le camion qu'en arrivant dessus. Là, vraiment, je *ne sais*
/confusion/

/confusion/

(30) *pas* ce qui se passait, c'est comme si je *ne reconnaissais pas*

(31) ce que je cherchais.

(32) Au retour, Céline était inquiète, elle n'arrêtait pas de me demander

(33) si ça allait. Elle me disait que je devrais me reposer un peu, que

/confusion/ */itératif/*

(34) la fatigue me fait *perdre la tête par moments*. C'est vrai que

/*confusion*/
(35) j'ai du *mal* à me souvenir de ses emplois du temps, mais aussi

(36) je trouve qu'elle change sans arrêt ce qui était prévu. D'ailleurs,

(37) j'en ai autant pour Jo qui prend des commandes pour un jour et

(38) décide tout d'un coup de faire la livraison un autre jour que celui

(39) qui était prévu. Alors s'ils s'y mettent tous, il n'est pas étonnant

/*confusion*/
(40) que je me *mélange* un peu les pédales.

Considering all of the verbs with the author as syntactic subject, we might identify a "false" active pathway or passive agency; "se perdre" (meaning "getting myself lost") was not a voluntary action, and Couturier, like other patients, experiences or endures events to a greater extent than enacting them. The same reasoning could be applied to "being tired", "struggling", etc.

As we have mentioned, Asp and de Villiers did not predict markers for loss or memory-linked deprivation. There is a good reason for this. Although we might expect to find equivalent expressions in all types of corpora containing discourse by early stage Alzheimer's patients, there are none in the following interview (Asp and de Villiers 2010, pp. 143–144), on a similar topic to Couturier's diary entry above. The genre of Couturier's text may have something to do with this, as the diarist distances herself rather more from her own mental state, using expressions such as *bribes* and *pointillés* to describe it.

There are several common points between the text above and that below. The text that follows is not a diary "monologue", but a dialogue recorded in a clinical setting, similar to extracts we studied previously.:

(1) IV: Can you tell me what happened that day.

(2) P: Well # uh it was uh just it it looked different to me,

(3) P: Like I mean # some of them streets # I uh see I been in that place for quite a few years # but I still # I just *couldn't*

quite get them streets figured out # so I just had to keep on walking # slow walking # and just try to figure out where # where I had to go # you see. But uh it took a good part of the day to get it figured out.

(4) IV: So what did you do to find your way back # did you ask someone or call?

(5) P: I did talk to the odd one # like you know # but they were uh kind of uh new too like the ones I were # was talking to and uh so they didn't really give me any help as far as the directions were that I wanted to go.

(6) IV: Umhum.

(7) P: I wasn't a bit worried about it.

(8) P: I just thought I'll keep on going till I find exactly where to go.

(9) IV: Do you know your address off by heart?

(10) P: Uh it's at uh # well # the one I'm staying at that's at her her uh her place here.

(11) IV: Can you tell me what the address is?

(12) P: Well that's uh # I I should have been able to.

(13) P: I have it written down # too. But ...

(14) CG: Yeah he has it in his pocket.

(15) P: Yeah ()

In this extract, the patient confirms that they felt lost ("I just couldn't quite get them streets figured out"), but, surprisingly, kept on going ("so I just had to keep on walking") without asking for help. That said, the patient still seems to be well aware of their impairment ("I should have been able to"). Linguistic expressions of disorientation are a clear indicator of spatial memory problems, as linguistic memory is simply an extension of this. It is likely that awareness of gaps in the memory disappears in the latest stages of the disease, along with the linguistic manifestations of these gaps, leaving only the expression of ill-defined confusion.

To conclude our presentation, we will analyze an extract which we consider to be one of the richest texts written by a person with Alzheimer's disease. This extract from Couturier's diary is a highly personal reflection on the disease, and on the author's life itself. With the exception of *entre les rêves et la réalité* (between dreams and reality), there are no confusion markers here; instead, we find an expression of anxiety (*crains* (fear), *angoisse* (anxiety)) resulting from the loss of her memories and, consequently, of her identity. There are several indications of imminent or existing /danger/; the use of substantives – some metaphorical, others literal – reflects an alternance between almost dreamlike moments, pauses and moments of full consciousness where her fears emerge: examples include *taureau* (bull) or *Minotaure.* There are two main aspects that feed into this anxiety:

1) First, the fact that the disease is neurodegenerative. The text features numerous combinations of /progressive/ x /iterative/ markers, including, in the first case, *toujours* (always), *de plus en plus* (more and more) (x2) and *petit à petit* (little by little), and in the second case, *reviens* (come back), *revenue* (note the presence of the verbal prefix) and *par moments* (by moments). Repetition and duration illustrate the inevitable nature of the loss, as do the future *j'aurai (oublié)* (I will have forgotten) and *en désespoir de cause* (in desperation).

2) Second, and perhaps the most poignant, reason for the author's anxiety is the idea of being doubly imprisoned, both internally and externally. In terms of internal imprisonment, we find lexemes relating to thought, dreams or nightmares (*cauchemar*, line 5), such as *labyrinthe* (x3), *fil d'Ariane* (thread, in the sense of a trail of crumbs) and *dédale* (maze); external claustration is expressed in every manifestation of absences that make the patient feel isolated from others or from herself.

These aspects are found in the final part of the text, where Couturier speaks about her personal situation. She compares this situation to that of another patient, and the pronoun "elle", used differentially, is a mark of isolation in its own right. The final paragraph of the text contains an unusual number of first-person markers; the *je*, evidently, and also possessive adjectives, including four instances of *mon* in just a few lines. Alongside these pronouns, we find an adverb, *personellement* (personally); a substantivating construction, *soi-même* (oneself); and adjectives, *intime* (intimate) and *seul* (alone, only). This over-representation finishes with the formulation *je me raconte mon histoire* (I'm telling myself my story), which features three first-person morphemes. In this way, Couturier seems to be attempting a form of self-saving, which is bound to fail.

/progressif/ /duratif/ /claustration/ /confusion/

(1)[8] "Tiens, j'en *reviens toujours* à un *labyrinthe* entre les *rêves* et la

/danger/

(2) réalité, avec en plus un *taureau* dans le coin, ça ressemble *de plus*

/progressif

/danger/ /échec/

(3) *en plus* à la légende grecque du *Minotaure*! *Si seulement* je

/claustration/

(4) pouvais, comme Thésée, avoir un *fil d'Ariane* pour pouvoir

/claustration/x /épreuve/ /claustration/

(5) *retrouver la sortie* dans ce *cauchemar*.

/danger/

(6) Je *crains* le jour où, même en relisant ces lignes, elles *ne me*

/échec/ /épreuve/

(7) *rappelleront plus rien*, car il m'arrive *d'avoir à* réfléchir

/épreuve/ x /échec/

(8) longuement pour *essayer de me souvenir* de certains événements

/échec/

(9) que j'ai relatés ici … par contre, je me souviens avoir *oublié*

(10) d'écrire certaines choses plusieurs jours après que cela soit

/claustration/

(11) arrivé… je vis à retardement. Quel *dédale*! et me voilà *revenue*

/itératif/

/claustration/

(12) dans mon *labyrinthe*. Si je suis l'architecte, mon fils Icare doit

8 The English version of the following text will not be presented alongside due to complexities with the layout. Any reader requiring the English translation can find it in Appendix 2 at the end of this book.

/danger/

(13) vite venir à mon *secours* avec ses ailes, avant que le monstre me

(14) dévore … Je n'aurais pas dû lire toutes ces légendes dans ma

(15) jeunesse: ça fait trop travailler mon imagination et nourrit mes

/danger/

(16) *angoisses.*

(17) Mais je ne suis pas la seule à avoir cette impression de

/claustration/

(18) *labyrinthe*: une malade américaine a écrit un livre que je viens

(19) de prendre à la bibliothèque, qui est son histoire; comme moi,

(20) elle a eu besoin d'écrire ce qu'elle vivait, car tout lui *échappait*
/échec/

/progressif/ */isolement/*

(21) *petit à petit*, mais *elle*, elle était soutenue et encouragée par son

(22) entourage pour écrire.

/isolement/ */inéluctable/*

(23) *Personnellement*, je le fais un peu en *désespoir de cause*: c'est

/isolement/ */isolement/* */isolement/*

(24) *mon* fil d'Ariane, *mon* vase d'expansion, *ma* soupape de

/isolement/ */isolement/* */isolement/*

(25) sécurité, le *seul* lien assez *intime* pour me rattacher à *mon* vécu

/isolement/

(26) (*personne* ne pourra me le raconter comme je l'ai ressenti quand

/inéluctable/ */échec/*

(27) je *l'aurai* oublié), et en plus, je *n'arrive pas* à m'exprimer

/échec/ /itératif/
(28) oralement comme je le voudrais. Ça *coince par moments*; je

/progressif/
(29) deviens *de plus en plus* émotive à ce sujet, donc, comme on n'est

/isolement/ /isolement/
(30) jamais mieux servi que par *soi-même*, *je me raconte mon* histoire".

It strikes us that, alongside the semantic features mentioned above, words relating to the lexical semantic field of memory are systematically associated with others belonging to the categories /try/, /difficulty/ or /failure/: for example, *essayer de me souvenir*, *ne me rappelleront plus rien* and *je l'aurai oublié* (in this final structure, *oublié* – forgotten – is an actual negation of memory). The verb *se souvenir* is used once in the affirmative, but ironically, with forgetting as the direct object: *je me souviens avoir oublié* (9). In some sense, Couturier disproves the idea that individuals with Alzheimer's disease "forget that they forget", and this increases her suffering.

In addition to interviews and diaries, we can gain an idea of the anxiety and the impression of claustration, or imprisonment, among patients through intimate moments captured using ongoing video monitoring. In the fascinating documentary *Strip-tease – la mémoire qui flanche*[9], we see a patient who is "wandering" for a few minutes and who, luckily, expresses her thoughts under her breath. This wandering is spatial as much as metaphorical: the individual walks around the garden, before returning upstairs to her house where her grandson is waiting (a fact of which she may or may not have been aware):

(1) P[10]: //Moi je reste xxx//C'est d'accord. xxxx//Ils sont tous partis//Qu'est-ce que je fais là?//# dans le jardin //Qu'est-ce que je fais dans # dans ton jardin?//Ouais, ouais//Il y a des fleurs

(1) P: //I'm staying xxx//Okay. xxxx//They've all gone//What am I doing here?//# in the garden //What am I doing in # in your garden?// Yeah, yeah/There are flowers everywhere//poor parents//Off we

9 *Strip-tease – la mémoire qui flanche*. Documentary extract, France 3, director: Barthélémy, program directors: Limon and Lamensch, producer: Fregosi [Online]. Available at: https://www.youtube.com/watch?v=p3Bd4iRxlqI (publié le 23/03/13), min. 12:05–14:00.

10 CG = caregiver (in this case, the patient's grandson); P = patient.

partout//pauvres parents//Passons (sortir)//Qu'est-ce que vous voulez que je fasse, moi?//Rien. xxx //Ouais//Y'a rien à bouffer//Catherine//Viens me chercher//Moi je meurs ici toute seule//Oui, je meurs ici//toute seule//Catheriiine//Oh viens à mon s'cours//Moi je veux pas rester ici//#Alors#	go (out)//What do you want me to do about it?//Nothing. xxx //Yeah//There's nothing to eat//Catherine//Come and get me//I'm dying here on my own//Yes, I'm dying here//on my own//Catheriiine//Oh come and help me//I don't want to stay here//#So#
(2) CG: Elle va venir Catherine, Mamie, tu sais?	(2) CG: Catherine's coming, Grandma, you know?
(3) P: Hein?	(3) P: Huh?
(4) CG: Catherine.	(4) CG: Catherine.
(5) P: Oui?	(5) P: Yes?
(6) CG: Elle vient.	(6) CG: She's coming.
(7) P: Elle vient ce soir?	(7) P: She's coming this evening?
(8) CG: Ouais.	(8) CG: Yeah.
(9) P: Mais # elle vient toute seule?	(9) P: But # she's coming on her own?
(10) CG: Avec Isabelle.	(10) CG: With Isabelle.
(11)P: //Avec Isabelle//Elles viennent toutes les deux?//	(11) P: //With Isabelle. They're both coming?//
(12) CG: Ouais.	(12) CG: Yeah.
(13) P: //Oh, nom d'une pipe//qu'est-ce qu'on va leur donner à manger?//((rire)) Attends, ça devrait être très sympa.//	(13) P: //Oh, golly//what are we going to give them to eat?/((laughs)) Wait, that should be real nice.//

This rare and extremely useful extract shows that the patient cannot remember her parents' death, despite having organized their funeral. There are clear lexical markers of anxiety – *meurs* (dying) x2, *s'cours* (help) and *pas rester ici* (don't want to stay here); the current time (all verbs are in the present tense) and place seem, strangely, to represent a danger, and are all described dysphorically: in the kitchen, there is nothing to eat (privation); the well-tended garden has flowers everywhere (perceived as a sort of invasion), leading to an exclamation about the patient's "poor parents". Other characteristic features are present, such as the feeling of isolation (*toute seule* x2) and uncertainty (*qu'est-ce que je fais là/dans ton jardin*); there is also a large number of movement verbs (*partis, sortir, fasse, viens, chercher*) and a fear of ceasing to move (*pas rester*). The word *rien*, moreover, relates to the possibility of movement (*que voulez-vous que je fasse, moi?*) which the patient does not think possible. The patient's daughter, whom she calls out for twice in the extract, will visit that evening; however, the patient is still concerned about the absence expressed in *rien: qu'est-ce qu'on va leur donner à manger?*

4.3. Analyzing patient discourse: perspectives

We have focused here on outlining two broad approaches to the discourse of patients with Alzheimer's disease: first, an eclectic quantitative analysis, and second, a qualitative analysis based on interpretative semantics. In both cases, schedules and strategies are still the very early stages of development. Everything now depends on linguists' abilities to create varied corpora of a useable size, and thus – indirectly – on the availability of funding from foundations and research organizations, a perennial problem in the humanities and social sciences.

4.3.1. *Recursion, circularity and sequences: finding meaning through analysis grids*

For example, further research could be focused on establishing grids of specific linguistic features for each stage of Alzheimer's disease. We have already discussed the features of early stages, as seen in interview transcripts or patients' autobiographies. To fully understand the importance of this work, consider the difference between the text below – where the speaker alternates between remembering and forgetting – and the extracts presented earlier:

(1) CG: //Donc toi tu étais sous-lieutenant dans l'armée//et tu es partie pour te marier et pour être infirmière//	(1) CG: /So you were a sub-lieutenant in the army // and you left to get married and become a nurse//
(2) P: //Voilà//	(2) P: //That's right//
(3) CG: //Voilà//J'crois qu'c'était un mois d'bateau//c'était énorme//	(3) CG: //Right//I think it was one month away by boat//that's huge//
(4) P: C'était? #	(4) P: Was it? #
(5) CG: //énorme// le temps de bateau//qu'il fallait compter pour aller jusqu'en Indochine//	(5) CG: //huge// the time you spent on a boat//to get to Indochina//
(6) P: //Ç'a été dur parce que j'étais seule//Robert # euh # il était pas tendre, Robert//	(6) P: //It was hard because I was alone//Robert # euh # he wasn't kind, Robert//
(7) CG: //Oui//c'était un (gu) xxx //c'était un vrai guerrier//	(7) CG: //Yes//he was a (fi) xxx //he was a real fighter//
(8) P: //J'sais pas s'il vit, s'il est mort//Dire que j'l'ai su à un moment, y'a pas longtemps, puis que c'est parti//	(8) P: //I don't know if he's alive or if he's dead//Hey, I knew that at one point, not long ago, then it left me//
(9) CG: Il est mort, oui.	(9) CG: He's dead, yes.
(10) P: Il est mort. #	(10) P: He's dead. #
(11) CG: //Oui, il était malade, tu sais//Il était malade//	(11) CG: //Yes, he was sick, you know//He was sick//
(12) P: Ah, oui. #	(12) P: Ah, yes. #
(13) CG: Il avait une leucémie #	(13) CG: He had leukemia #
(14) P: Ah, oui.	(14) P: Ah, yes.
(15) CG: //Et la guerre d'Indochine, ça, tu t'en souviens un peu//de tous ces jeunes?	(15) CG: //And the war in Indochina, that, can you remember all those young men?
(16) P: Oui// beaucoup de choses m'échappent//mais #	(16) P: Yes// I forget a lot of things//but #

(17) CG: C'était la boucherie quand même	(17) CG: It was a real blood-bath
(18) P: //La guerre d'Indochine, c'était très dur//très dur//Ah, oui//	(18) P: //The war in Indochina, it was very hard//very hard//Ah, yes//
(19) CG: //Et en tant qu'infirmière, tu as dû en voir de toutes les couleurs?//	(19) CG: //And as a nurse, you must have seen a lot?//
(20) P: //Ah, oui//alors ça, ben, c'est bien//parce que ça te mobilise, ça t'oblige//	(20) P: //Ah, yes//so that, well, it's good, because it motivates you, it makes you//
(21) CG: … à faire des choses.	(21) CG: … do things.
(22) P: //Tu//tu demandes rien//Mme B., là-bas//Hop, j'cours//ah oui, j'ai couru # quand même//	(22) P: //You//you don't ask questions//Mrs. B., over there//Off I run//ah yes, I ran # that I did//
(23) CG: //Oui//Beaucoup couru, oui// ^^	(23) CG: //Yes//You ran a lot, yes// ^^
(24) P: //Tu vois//peu de chose, pouf! ça r'vient//	(24) P: //See//not much, and abracadabra, it comes back//
(25) CG: //Ca (t'ém) xxx //c'est très xxx//t'as beaucoup d'émotion quand tu repenses à ça?//	(25) CG: //It (bo) xxx //it's very xxx//does it make you very emotional to think back on that?//
(26) P: //Oui, c'était très dur//dur, cette guerre d'Indochine//#	(26) P: //Yes, it was very hard//hard, the war in Indochina//#
(27) CG: Ah, oui …	(27) CG: Ah, yes …
(28) P: Puis j'étais toute seule …[11]	(28) P: And I was all alone …

There are several interesting features in this extract. First, the patient clearly takes pleasure in remembering: *pouf! Ca revient*. The patient's daughter helps her in this respect, asking her mother about her past and recalling details of the general setting; this type of information, while familiar to family members, is rarely accessible to medical professionals. In

11 CG = caregiver (in this case, the patient's daughter); P = patient. *Strip-tease – la mémoire qui flanche*. Documentary extract, *Op. cit.* [Online], min. 5:58–7:38.

this way, the caregiver creates the conditions that make this experience possible. It is highly likely that the act of remembering, particularly when mixed with pleasure, will delay both the symptoms and the self-abandonment and isolation we have been talking about. These sequences where the patient remembers the distant past seem to function as "oases" of safety in communication, particularly in the context of dialogue; alone with a loved one, the patient walks through a set of information that is known or, at least, is still accessible.

But outside of these reassuring "oases" of remembering, danger lies in wait, particularly in the later stages of the disease, as we see here. The second thing we note is a type of circular functioning, something all caregivers observe. On several occasions in the documentary, the patient asks her daughter and grandson where her husband is, even though he has been dead for several years and she organized his funeral herself. She wonders what has become of him, or she wants to call him, not knowing where he lives, or she finds it strange not to have seen him recently; this can be tiring and frustrating for family members. The middle section of the extract above concerns the patient's husband; the patient's daughter tells her that he died of leukemia, but there is no doubt that the question will be asked again the next day. Another example of this looping phenomenon can be seen in the repetition of certain ideas: *parce que j'étais seule* (6) is mirrored by *j'étais toute seule* (27). Every caregiver knows how hard it can be to break the cycle once a patient is trapped. Changing the subject does not help. It seems as though there was an independence of sequences, with one theme at the center, in the memory of the patients. If a theme is not resolved – which it never is, since the patient cannot encode the recent past – it seems to remain the patient's sole concern; since this theme is often the fate of a loved one, the uncertainty becomes unbearable, and this explains the circularity of their discourse.

Note that this process of returning over and over to the same subject is a recurring feature in absurdist literature. Take the following example, from *Incidences* by the Russian author Daniil Kharms, which we have analyzed in detail elsewhere (2012, p. 155–158):

My dear Nicandre Andreïevitch...

> My dear Nicandre Andreïevitch, I received your letter, I realized right away that it was from you. At first, I thought it

> might not have been from you, but as soon as I opened the envelope, I knew right away that it was from you, even though I'd almost thought that it wasn't from you. I'm very happy to know that you've been married for some time, because when a person marries the one he wanted to marry, that means that he got what he wanted. That's why I'm so happy that you married the one you wanted to marry, because you finally got what you wanted. Yesterday, when I got your letter, I knew right away that it was from you: then I thought that maybe it wasn't from you, but once it was open, I knew that it was a letter from you. You were right to write me. You hadn't written me for a long time, but then, you wrote me, even though you wrote me before, before you stopped writing me for a while. As soon as I got your letter, I thought it was from you, then I was very happy to find out that you're married, because when someone wants to marry, he absolutely must marry …

The text goes on for another page, following the same *modus operandi*; the absurdity here lies in the construction of the text, based on a rather odd repetition of identical ideas. The author of the letter notes, over and over again, their hesitation concerning the identity of the correspondent, then their joy at knowing the correspondent is married, then their hesitation concerning the correspondent's identity and so on. The process of repeating the same ideas over and over again prevents the text from progressing. In the literature, this lack of movement often reflects a character's malaise, despite the letter writer's evident desire to remain alive and continue to "move", via his or her letters. Kharms breaks the law of communication according to which a speaker cannot, or should not, repeat the same thoughts *ad infinitum*, and/or produce a text which is, literally, nonsensical.

Yet the text follows – albeit in caricatural form – a "natural" pattern observed in dementia patients, who are unable to record and assimilate finished conversation sequences. The evident malaise of characters in absurdist literature, such as the author of this letter, is troublingly similar to the distress we find in patients. The final consequence of this complete inability to move forward is isolation and what Ploton (2004, p. 94) refers to as *fonctionnement mental de secours*: individuals with ATD no longer learn, whether from their own experience or from others, and, as we have seen, this results in a bi-fold isolation both from society and from the self.

In these texts, in spite of visible changes, the bases of communication are still partially retained. The same cannot be said of patients' speech in the final stages of the disease. To study these changes in depth would require a large corpus of interviews such as those recorded by Alois Alzheimer himself. Alzheimer recorded the following conversation with a patient named Auguste Deter, a few years before her death (Maurer and Maurer 1999, pp. 18–19):

29 November 1901

> "How much does a pound of meat cost? — Seventy. — Seventy what? — I don't know. — And a bun? — Three pfennigs. — What's 2 x 3? — 6 — 9 x 7? — 63 — 12 x 19? — 27 — 6 x 8? — 48 — If you buy six eggs at seven pfennigs each, how much will it cost? — You poach them. — What street do you live on here? — I can tell you, I just need a moment. — What did I just ask you? — But of course, it's Frankfurt-am-Main. — What street do you live on? — I can tell you that, Waldemarstrasse — no, no … — When did you get married? — I don't know for the moment, the woman lives in the same hallway. — What woman? — The woman, where we live. Sometimes she cries out: 'Frau Hensler! Frau Hensler! Frau Hensler!' She lives here, on the floor below".

At this stage in the disease, there is a visible difference with the texts studied previously: here, even the chains of reference are broken, and there are no longer any anaphoric links. There is no identifiable logical connection between "When did you get married?" and the sentence which comes almost directly after it, "the woman lives in the same hallways"; nor is there an obvious connection between "how much will it cost?" and Frau Deter's response, "You poach them".

Unfortunately, there are no sufficiently large corpora of this type. Considerable time would be needed to collect and annotate the data; furthermore, researchers would need to have access to the patients themselves and obtain their agreement, that of their families and that of the medical personnel. This is not impossible, but can be difficult for researchers from outside the medical field. Linguists – who are visibly not medical professionals – may be treated with some suspicion when they appear to be interested in patients' speech, and patients themselves may have other concerns and other desires than to see their language dissected.

Linguistic analyses could also help medical staff and caregivers to communicate more effectively with patients, for example by avoiding embedded subordinates and by focusing on a single constant theme, rather than on a hyper-theme that is developed through a progression involving derived themes. The speed of the speech flow is also important: this observation is nothing new to experienced caregivers. Anne L., who is her husband's caregiver, states, to camera and in front of her husband[12]:

//Alors c'est toujours difficile au jour le jour//de, de plus en plus difficile de se comprendre//parce que ##//parce que tu ne trouves pas tes mots //((rire)) et parce que si je fais des phrases trop compliquées, ou si je parle trop vite//((rire)) ça ne passe pas non plus //#il y a beaucoup d'intérferences dans les communications maintenant.	//So it's always hard from day to day//it's getting harder to understand one another//because ##//because you can't find your words //((laughs)) and because if my sentences are too complicated or if I talk too fast//((laughs)) it doesn't work either //#there's a lot of interference in our communication now.

This suggestion is nothing new: Asp and de Villiers recommend the same form of patient support, following initial recommendations made by Kempler *et al.* The issue has become an area of research in its own right in the United States and a clearer picture is forming regarding the strategies to be adopted. Several criteria related to memory impairment, most of which have already been mentioned here, are frequently observed; paraphrasing and repetition are likely to be used by Alzheimer's patients to counterbalance their encoding difficulties, and they are likely to avoid complex syntactic constructions. However, Kempler *et al.* (1998) note that speech speed does not appear to affect all patients in the same way: the greater the deficit in working memory – or, put differently, the more advanced the disease is – the less a patient will benefit from talking slowly. This is because slower speech means that the patient must maintain concentration for longer, and at this point, talking slowly can even hinder communications. These observations should be taken into account in establishing a scale for different ages and stages of impairment linked to Alzheimer's disease; they may also be used in

12 *Patients jeunes et aidants : leur histoire face à la maladie d'Alzheimer*. Documentary, Ligue Alzheimer ASBL [Online]. Available at https://www.youtube.com/watch?v=9tZT5sxCA7E&t=333s (published May 22, 2017), min. 9:15–9:37.

the development of linguistic strategies designed to improve communication with patients, thus reducing isolation.

4.3.2. *A multi-channel approach*

In this chapter[13], we have focused on the linguistic aspect of patient speech alone; we are well aware that this, too, is not entirely sufficient, given the insistence on a multi-channel approach in the field of interaction analysis. Several factors, of which language is only one, are at play in every conversation; gestures, tone and mimic, among other things, are crucial, and far from being a secondary level to consider in addition to language, non-verbal aspects can even be the most crucial part of a message, forming the central framework around which speech turns are constructed. This section aims to go some way to filling this gap in our work.

For example, the three following passages[14], transcribed from an audio recording, do not do justice to the interaction taken as a whole. Three male patients (noted P1-2-3 below), all of whom have been diagnosed with Alzheimer's disease, are asked about their former professions. Crucially, in this case, their wives are also present (W1-2-3); the looks and physical contact exchanged over the course of the interview appear particularly significant. The gestures made by individuals are also essential. The transcription below is annotated to show verbal hesitations, but watching the video, we see that the speakers wring their hands in a way that suggests they are having difficulties. The individuals speak very slowly, as if they were giving themselves the time to find their words. Remembering requires a significant effort and relies, in this case, on the contribution of the patient's wife, who supplies the missing expression *objecteur de conscience*:

(1) P1: //J'ai fait des é#des études de médecine//et je suis venu pour faire des//## un travail en tant que ## Euh.//Je ne trouve#//	(1) P1: //I stu#studied medicine//and I can to do//## to work as ## Uh.//I can't find#//

13 Much of this work builds on a conversational analysis course given at Masaryk University in Brno in 2019 and owes a good deal to discussions with students, notably Walová.

14 *Patients jeunes et aidants: leur histoire face à la maladie d'Alzheimer*. Documentary, Ligue Alzheimer ASBL [Online]. Available at: https://www.youtube.com/watch?v=9tZT5sxCA7E&t=333s (published May 22, 2017), min. 2:20–3:21.

(2) W1: //Objecteur de conscience//	(2) W1: //Conscientious objector//
(3) P1: //Objecteur de conscience// Maintenant, ça n'existe//#ça n'existe plus//	(3) P1: //Conscientious objector// Now, that doesn't exist//#that doesn't exist anymore//

In the second extract, transcribed below, the patient's eyes move, from apparently looking into space to seeking approval or assistance from his wife, who is called upon even more than her counterpart in the previous extract. The patient, slouched in his chair, barely responds to the camera, simply stammering a few words while looking for approval from his wife. Furthermore, it is the patient's wife who gives him the answers in a firm and doubtless somewhat irritated tone:

(1) P2: //Je faisais de la#de la #//hmm ((petit rire adressé à la conjointe))//	(1) P2: //I did#I did//hmm ((short laugh addressed to wife))//
(2) W2: //Il était#Tu as été professeur d'abord//	(2) W2: //He was#You were a teacher first//
(3) P2: //D'abord#ah.ah.#Oui//	(3) P2: //First#ah.ah.#Yes//
(4) W2: //De?//	(4) W2: //Of what?//
(5) P2: //Oui. De, d'électricité//et de #de mécanique##Euh.//	(5) P2: //Yes. Of, of electricity//and of #of mechanics##Uh.//
(6) W2: //Et puis après#//	(6) W2: //And then after#//
(7) P2: //Et puis on m'a demandé une charge de//##de##de###//	(7) P2: //And then they asked me to take a position as an//##an##an###//
(8) W2: //d'animateur//	(8) W2: //activity leader//
(9) P2: //d'animateur xxxxxxx//	(9) P2: //an activity leader xxxxxxx//

As we see here, the healthy partner becomes the custodian of the couple's memory as the other progresses through the stages of Alzheimer's disease. Both partners are aware of this, and it represents an additional burden on "family" caregivers, on top of more practical obligations. This necessarily has an effect on relationships. The third patient is also ill at ease in front of the camera, and constantly appears to seek support from his wife. In addition to eye movements, the movements he makes with his right hand are all destined to check, with his wife, whether what he says is right; the wife, sitting with her arms crossed, does not seem particularly happy in her role:

(1) P3: //J'ai été dans l'équipe seniors Belges//J'ai joué contre la Hollande//dans le nord de la France//et j'ai raté le #### ((légère tape sur le bras de son épouse))//à## ((signe de tête vers le haut)) ###au championnat d'Europe.	(1) P3: //I was on the Belgian senior team//I played against Holland// in northern France//and I missed the #### ((light tap on wife's arm))//to## ((indicates upward with head)) ###in the European championships.
(2) W3: //Ah oui//En Autriche?//	(2) W3: //Ah yes//In Austria?//
(3) P3: //Oui//	(3) P3: //Yes//
(4) W3: //Oui//T'étais presque qualifié//	(4) W3: //Yes//You almost qualified//
(5) P3: //Exactement//Ouais//	(5) P3: //Exactly//Yes//

This situation is all the sadder as, clearly, this memory marks one of the highlights of the patient's life story: he almost had an international sports career. However, while he seems proud of the fact and wants to talk about it, the patient is unable to access the whole memory.

Returning to the documentary cited earlier, *La Mémoire qui Flanche*, we encounter another patient whose body language is particularly telling. She never seems to look directly at the person to whom she is speaking, and her eyes seem to glaze over; furthermore, the patient has a number of tics which could be seen as signs of agitation, indecision and uncertainty. In the following extract[15], patient "M" is out walking in the street and encounters her grandson "PF". We have attempted to annotate interesting elements of body language in parentheses:

[vehicle noise]

(1) M: ((soupir)) Je suis tellement # moche, mon pauvre ami# ((passage de la langue sur toute la	(1) M: ((sigh)) I'm so # ugly, my poor friend# ((runs tongue across bottom lip)) # It's not worth #

15 *Strip-tease – la mémoire qui flanche*. Documentary extract. *Op. cit.*, min. 8:58–9:30.

lèvre inférieure)) # Ça n'mérite pas # un p'tit coup d'œil là.	looking at.
(2) PF: //Comment ça va, mamie?//	(2) PF: //How are you, grandma?//
(3) M: //Oui, mais#//	(3) M: //Yes, but#//
(4) PF: //Ça va?//	(4) PF: //You okay?//
(5) M: //# Oui, mais#//	(5) M: //# Yes, but#//
(6) PF: //T'étais où?//	(6) PF: //Where were you?//
(7) M: //Hein?//	(7) M: //Huh?//
(8) PF: //T'étais où?//	(8) PF: //Where were you?//
(9) M: //Ben, j'sais pas//où les gens m'envoient//Je vais # et j'reviens # (rire) Écoute …	(9) M: //Well, I don't know //where people send me//I come # and go # (laughs) Listen …
(10) PF: //Tu t'souviens de moi, mamie?//	(10) PF: //Do you remember me, grandma?//
(11) M: //Ah oui ((très certaine)//	(11) M: //Ah yes ((very sure)//
(12) PF: //Tu sais comment je m'appelle?//	(12) PF: //Do you know my name?//
(13) M: //Ou.ouiii ((ton hésitant))// # Oui ((moins hésitante mais peu sûre, rire)). //Si, je me souviens bien de toi ###((regard un peu triste jeté à part))//	(13) M: //Y.yeees ((hesitant tone))// # Yes ((less hesitant but unsure, laughs)). //Yes, I remember you well ###((slightly sad gaze off to one side))//

The most striking elements are perhaps the way the patient runs her tongue across her bottom lip, and the sad/defensive laughs. The woman is wandering, and clearly suffers from agnosia; she attempts to compensate for gaps in her memory by pretending, verbally, to know and recognize things, but her corporal semiology – notably her gaze – gives away the truth. Evidently, even as linguists, it is crucial for us to take account of the other channels of communication at work in every interaction.

4.3.3. *Confabulations: "from minor distortion to bizarre tales"*[16]

In our view, we also need to consider the issue of *confabulations*, as discussed by Barbizet and Truscelli, but from a linguistic perspective. We have already discussed Barbizet's "Lion's Tale" test at some length; one of the main objectives of this test is, in fact, to assess confabulation (as we see from the sub-title of Barbizet's article, *Considérations sur la fabulation*). Furthermore, confabulation is intrinsically linked to amnesia, as Schnider noted (2008, p. 147): "[…] nobody seriously doubts that severe confabulation after brain damage requires the presence of amnesia".

As linguistic materials, confabulations are truly fascinating; however, as yet, few linguists have exploited this material. Neurologists, on the other hand, realized the importance of confabulations early on, although there is no consensus over how to address them. Barbizet's scoring grid only leaves one space for confabulations, at the very end of the sheet; this result has no influence on the psychometric assessment score.

That said, our understanding now goes much further than Delay's notion (cited by Barbizet and Truscelli 1965, p. 1693) that "a confabulation is not a story, it is a form of delirium, in which each person expresses their own thought following the passage of received information, on the basis of prior acquisitions". Authors such as Dalla Barba have made considerable contributions to the subject, although the basic premises of neuropsychology are never questioned, as in all of the other works discussed here. In terms of assessment, patients are generally given a list of items (words or pictures) to use in narrating an event. Current work on confabulation focuses on identifying patterns, mechanisms or models of confabulation, ways in which these might be triggered, which types of cerebral lesion tend to produce them, and, consequently, which types of patients are most likely to confabulate. Barbizet and Truscelli paid particular attention to these two final points, notably in patients with alcoholic Korsakoff syndrome, an amnestic disorder identified by the eponymous Russian neuropsychiatrist in the late 19th Century.

For us to speak of confabulation, a patient's utterances must be both involuntarily and unconsciously false, that is, must not be deliberate fabrications; the patient is speaking in good faith, as Dalla Barba states (1993,

16 Expression taken from Dalla Barba (1993, p. 567).

p. 567): "Confabulation is a not well understood symptom, observable in amnesic patients who are unaware of their memory deficit. It consists of both actions and verbal statements that are unintentionally incongruous to the patient's history, background, and present situation". While actions do not concern us here, verbal statements are evidently relevant to our discussion. The aim is thus to find out why individuals confabulate; no answer can be found without a sufficient collection of instances of confabulation, which must be analyzed to identify both their nature and contents. As is often the case in neuropsychology, there is a risk of collecting only artificial, or artificially triggered, cases. Lexical tests, of the type presented here, are unlikely to produce large amounts of confabulation; better results can be obtained from recordings of anamneses or, better still, summaries of stories, as in the case of the Lion's Tale test. It is evident that longer interviews are also valuable; however, the time required to carry out and analyze these interviews is prohibitive, given the sizeable workload of all neuropsychologists.

The bulk of recorded confabulations are responses to simple questions or, at best, utterances in the context of semi-guided interviews. While Dalla Barba (1993, pp. 571–572) does not provide the full set of questions used, he presents a relatively detailed outline:

> "(1) Twenty questions probing Personal Semantic Memory (age, date of birth, current address, number of children, etc.).
>
> (2) Fifteen questions probing Episodic Memory.
>
> (3) Ten questions probing Orientation in Time and Place.
>
> (4) Fifteen questions probing General Semantic Memory (knowledge of famous facts and famous people, past and present).
>
> (5) Fifteen questions probing Linguistic Semantic Memory. Items 16 to 30 of the WAIS vocabulary subtest were selected for a word definition task.
>
> (6) Ten 'I don't know Semantic' questions, i.e., questions so constructed as to be likely to be appropriately responded 'I don't know' by normal subjects (e.g., 'What did Marilyn Monroe's father do?').

> (7) Ten 'I don't know' questions relating to episodic memory, for example 'Do you remember what you were doing on March 13, 1985?'"

In terms of identifying confabulations, it is easier to identify a fabricated answer to a type-6 question than to a question designed to evaluate personal memory; nevertheless, certain biographical data is accessible and can be measured objectively, and other details can be verified with family members. Control data is crucial in order to determine whether or not an utterance of this type is confabulated.

Several authors suggest that a clear distinction should be made between confabulation and *intrusion*, in which one element replaces another in a memory test. Intrusion is necessarily artificial, as it is triggered by the question context; in our view, it is therefore out of place even in the first category of confabulations defined by Schnider (2008, pp. 73–74). On the contrary, the bulk of instances of confabulation used by Dalla Barba were obtained in the course of tests of this type. All things considered, there is no self-evident typology for confabulations. Schnider suggests the following four categories:

> "1) Intrusions or simple provoked confabulations in memory tests.
>
> 2) Momentary confabulations upon some form of incitement, be this questions or the circumstance of a discussion. False statements can range from a simple response to a question to elaborate confabulations. Momentary confabulations are probably not a unitary disorder.
>
> 3) Fantastic confabulations, which have no basis in reality and are intrinsically nonsensical and illogical.
>
> 4) Behaviorally spontaneous confabulations, that is, confabulations reflecting a confusion of reality, which underlies currently inappropriate behavior and statements".

The author does not formulate the criteria used to determine these categories, but we note at least three main themes:

– the *source* that triggers confabulation (tests for category 1, questions or discussions for 2, and no source at all – i.e. spontaneous confabulation – in 4);

– the *nature of the content* (intrinsically absurd in category 3, or not absurd, in others, or inappropriate to the context, as in 4);

– the *momentary* (2) or, perhaps, more *durable* (3, 4) nature of the phenomenon.

Inversely, *falseness* is present in some way in all of the categories. Category 3 is problematic from a linguistic perspective, as we know from our own work in this area (see, notably, (Cusimano 2012, 2015)). Neither absurdity (nonsense) or apparent illogicality should be taken lightly; man, as an "interpreting machine", inevitably gives meaning even to things which are, at first glance, devoid of meaning. Furthermore, the absence of meaning is often, in practice, limited to one interpretative step and can be resolved by the addition of further details. Even though linguists and neuropsychologists were able to reach agreement concerning what, exactly, constitutes a nonsensical utterance, it would still be wise to examine each case on an individual basis, something which we shall attempt to do below.

In terms of explanations for the phenomenon, Malaguarnera's definition of confabulation (2016, p. 93) is a helpful starting point. This definition is unique in that there is no distinction made between defining and explanatory aspects: "The term *confabulation*, synonymous with *fabulation*, refers to the production of imaginary inventions destined to mask reality". In passing, it is not at all certain that confabulating patients actually aim to conceal part of reality, given that their invention is involuntary. The notion of a "replacement reality" sets this definition apart from other definitions based on the idea of gap-filling, which was, for a long time, the dominant belief and which still has its supporters. Authors such as Bonhoeffer, as early as 1901, suggested that confabulation is literally a means of filling in the gaps left by lost information; although Bonhoeffer was writing about momentary lacunas, other researchers extended this explanation to the domain of fantastical inventions. As we have already seen in passing, confabulating patients make no difference between what is, or has been, real, and what they themselves have invented. Some have suggested that there is an emotional element to this need to fill in gaps in the memory, although this does not sit easily with the notion that patients are not aware of their impairments; in practice, even individuals who appear to be entirely unaware of their amnesia are never completely ignorant, and all, in their own ways, develop verbal techniques to work around problems that may arise as a result of the gaps in their memory. In this way, they make at least some effort to avoid embarrassing situations. Schnider (2008, p. 207), writing about patients with severe forms of amnesia, is a firm proponent of this view:

> "They were fully aware of their amnesia. They would not insist on the veracity of their momentary confabulations. Clearly, they had developed a strategy to avoid being embarrassed by questions they could not answer. Occasionally, patients seem to resort to gap-filling as a way of dealing with what they recognize to be implausible scenarios or inconsistencies in their knowledge. Notwithstanding such single cases, in which gap-filling appears to be the obvious explanation for momentary confabulations, controlled data from groups of patients have established that gap-filling is no generally valid account for confabulation".

Note that this hypothesis does not offer any explanation why certain confabulations occur even when there is no gap that needs filling, appearing spontaneously and not in response to any visible need. Evidently, the gap-filling theory raises as many questions as it solves.

There are several different approaches to the subject of confabulations, all building on Dalla Barba and Schnider's seminal work in the area (to the best of our knowledge, Schinder's book is the only one fully devoted to this subject). Cole *et al.* (2014, p. 208) took an original, while relevant, approach, focusing on the connections between confabulation about past events and the ability to "confabulate the future":

> "A recent focus in human memory research is the overlap between cognitive processes used in retrieving past events from one's personal life and the construction of imagined future events. Neuroimaging and behavioral data suggest that imagining and remembering draw upon similar brain regions and mental processes [...]. In the current paper we examine these issues from a neuropsychological viewpoint. In short, does a patient who re-constructs implausible memories of non-existent past events also show a similar generation of implausible future events?"

Cole *et al.* carried out a qualitative study focusing on one single patient, a police officer who began to confabulate excessively and with no inhibition after suffering a stroke, becoming what the authors refer to as a "persistent fabulator". This case was made all the more interesting by the fact that the officer's wife was still alive, meaning that the researchers were able to verify biographical information. The tests used by Cole *et al.* were also original. Using two series of eight concrete nouns, the patient – MW – was asked to recall personal events and to imagine possible future events. These future events were assessed on a scale from 1 to 7 using a two-step process. In the first step, the patient was asked to estimate the plausibility of the events, and in a second step, five control subjects were asked to give a similar estimation. The problems with this procedure seem evident: objective assessment would be impossible even if 100 control subjects were used, a fact that no number of mathematical formulae – of which many are given in the article – can mask. The results are mixed, but appear to confirm the connection between past and future confabulation, to the exclusion of the present, which seems to respond to other processes. Furthermore, the results obtained from this patient echo those obtained by other authors studying patients with Alzheimer's disease, as Cole *et al.* indicate:

> "Also, in a recent study of Alzheimer's patients, past and future confabulations correlated positively, whereas present orientation failed to correlate with either (Lee *et al.* 2007). Thus, there may be differences between present orientation and other forms of subjective time which involve mental time travel (Tulving 1985)".

In terms of the content of confabulations, MW had a striking propensity to see himself as powerful and possessive, suggesting a change in his self-image. The authors of the study state that this, too, corresponds to other findings on the connection between confabulatory activity and changes in habits and desires.

There is a wealth of available literature on the topic of confabulations, and it would be difficult to provide a thorough overview here; however, this brief state of the art is sufficient for our purposes. In the following section, we will consider the confabulation phenomenon in connection with Barbizet's Lion test, aiming to gain an idea of the variety of confabulations which patients produce in this context and to provide some idea of the underlying logic. This approach is similar to one proposed by Dalla Barba himself (1993, p. 580): "We have further argued that the confabulation content must be considered in terms of its semantic structure and depends on the relative level of integrity of semantic knowledge". As linguists, our idea of semantic knowledge is somewhat different, but the fact that medical researchers have highlighted this aspect provides a clear incentive to continue our research in this direction.

Patients have produced an impressive variety of confabulations in response to the Lion test. The first attempt at classification was made by Barbizet and Truscelli, and – despite the age of the classification and the fact that the authors were not linguists – still merits closer examination. Given the impossibility of correlating etiological and lesional factors with the confabulations themselves[17], the authors chose to focus on a detailed analysis of patients' inventions and of "deviations" from the story, including ((Barbizet and Truscelli 1965, p. 1689–1690), somewhat modified due to the authors' decidedly un-academic style of presentation):

1) changes in the animals or character names/functions;

2) modifications to "scenic movements";

3) confusion regarding characters' feelings;

4) modifications to the conclusion or variants in interpretation;

5) loss of the general meaning of the story.

There is an interesting distinction to be made between the following:

– *Substitutions*: cases where the patient replaces the lion with a bear, a tiger or even a wolf; replaces the keeper with a tamer (and the zoo with a circus); replaces the woman with a grandmother (all category 1); cases in which the lion steals the child or where the lion is freed rather than escaping (category 2); cases where, at the end of the story, the mother runs home (category 4).

– *Interpretations*: these are harder to analyze. There is a further distinction to make between *plausible* and *non-plausible* (or less plausible) explanations. For example, the statement that the woman is a mother is

17 "Let us state from the outset that this study of etiological and lesional factors does not go any way to explaining the production of confabulations" (Barbizet and Truscelli 1965, p. 1690).

doubtless the result of interpretation; however, in practice, it is quite likely that a woman carrying a small child in her arms is the child's mother. We could argue that this type of statement is not really a confabulation at all.

In our view, all other interpretations are the key to understanding confabulation and constitute the most interesting cases. Certain "looser" confabulations seem to be the product of unlimited interpretative activity, in which the patient rewrites, extends or enriches the story with absolute freedom. Most of the time, it would be wrong to say that these inventions are illogical, as they still obey linguistic norms, and still respect the hermeneutic principle of determining the local from the global.

To take just one example, Barbizet and Truscelli appear surprised by some of their observations, betraying a certain naiveté (Barbizet and Truscelli 1965, p. 1690): "What surprises us most is the frequency with which patients extract a moral from the story". It seems evident to us that, through semantic actualization, patients recognize features of the *fable* genre (rightly or wrongly) in the text; fables always finish with a moral. When a patient suggests that "the lion's owner, well, he should have to pay a big fine, and that way other people who have animals should think twice", he or she is simply producing an epilogue, which the text – seen as a fable – seems to invite.

Evidently, healthy subjects do not give in to the interpretative inclination present in every speaker; they remain in control of their speech and are well aware that this is not what is expected of them. However, it is clear that patients follow a certain logic, perhaps the same logic that explains the replacement of the lion or other characters in the story by other animals encountered in fables they already know. This point is also made by Barbizet and Truscelli (1965, p. 1692): "Other [patients] return to the realm of their own childhood, where the Lion's Tale triggers memories of fables or nursery rhymes: 'Ah yes, the story of the Wolf and the Lamb, or the Crow and the Fox'". This observation also offers a likely explanation for the ways in which patients anthropomorphize the lion (e.g. in "the lion, touched by maternal feeling ..." or "the lion saw a glimpse of understanding in the eyes of the mother ..."): in fables, animals are doted with reason and the same feelings as men. Patients also tend to improvise on the theme of character sentiments ("she begged the lion, she looked the lion in the eyes despite her fear"), likely to counteract the perceived "dryness" of the tale, which shares many other features with classic fables.

This last idea is purely hypothetical, but its validation would come as no surprise to most analysts. The textual genre is inherently interpretative, and the very act of interpretation is an inextricable part of the process. As we noted earlier, our updated version of the Lion test is, at the time of going to print, being trialled[18] with Alzheimer's patients. The test has already given rise to a considerable volume of fabulations, just as it did for Barbizet and Truscelli: for one patient, the lion became a "doggy" and the child "a little boy called Pierre"; for another, the woman is transformed into a "fairy dressed in blue", straight out of a fairy tale. The extract below[19] followed directly on from a first reading of the Lion's Tale:

(1) ST: //Alors dites-moi//est-ce que vous pouvez#euh##//essayer de retrouver la trame de l'histoire …//	(1) ST: //So tell me//can you#uh##/try to pick up the thread of the story …//
(2) P: //Une petite histoire merveilleuse//#c'est un conte de fée//#xxx la fille en bleu c'est une fée qui est habillée en bleu//	(2) P: //A marvellous little story//#it's a fairy tale//#xxx the girl in blue is a fairy dressed in blue//
(3) ST: //Oui#//	(3) ST: //Yes#//
(4) P: //C'est, c'est joli//	(4) P: //It's, it's sweet//

For this patient, the fact that the woman was "dressed in blue" left a strong trace in the memory and seems more important than we might have thought; perhaps it is the color that connects the woman to a fairy in the patient's mind. On the contrary, it is not surprising that, as the authors predicted, the later recall by the same patient ends with a moral, which is, moreover, clearly indicated: "Moral of the story: animals remember things because she was a fairy". In response to the speech therapist's extended "ah", the patient adds: "well, yes, he couldn't let it go like that", indicating that the specific and characteristic logical connections of the identified genre were preserved, despite their somewhat idiosyncratic application in this case. Listening to, or reading, a fairy tale, children are never surprised by the magic powers of a fairy, witch or even object; once established, the exceptional

18 Marie Montillot, a speech therapy student at the *École d'orthophonie* in Besançon, France, is working on a new scoring system for the test.

19 Full transcriptions of passages relating to the Lion's Tale are provided in the appendix.

characteristics of an object or person is unquestioned, and remains established for the remainder of the story.

As we see in this case, the patient has immediately established, in order, that:

1) The story is a fairy tale.

2) That a story of this sort must include a moral, which he, personally, has understood.

3) That the identity of the woman – as a fairy – is key to the story.

The patient did not "lose" this interpretation/confabulation, unlike so many other elements and details of the story; the same elements were present in the delayed recall test.

Other category 5 confabulations – that is, those which are considerably different from the original tale – are much harder to analyze. Barbizet and Truscelli themselves were perplexed by the phenomenon: "One can only wonder what mechanisms are at work for subjects to come up with such surprising responses to a simple tale with only 22 items". At this stage of our research, we are unable to provide any definitive answers; however, in our opinion, these confabulations could be analyzed using the same angle of approach as before. We suggest that the patient is often telling another story, in which the genre parameter is sometimes the only element which is retained. For example, "Story of a wolf or a tiger, he ate someone"; "Story of a wolf, a little boy, a kid, right, that he chucked on the ground" and "The hare with the woman who got her head cut off". The patients making these utterances at least recognize the text as a "story", and the replacement of the lion by animals such as a wolf or a hare seems to suggest an assimilation with other short stories, such as fables or fairy tales. However, this hypothesis still needs to be validated using a greater number of confabulations.

These more "distant" evocations of the story frequently display a significant quantitative reduction, although this is not systematic; according to Barbizet and Truscelli (1965, p. 1690), "although a reduction of the story seems to be the norm, it is not absolute, and the occasional patient will confabulate at considerable length". This propensity for extravagant confabulation should perhaps be considered in relation to the patient's capacity for general communication: for example, we know that certain late-stage Alzheimer's patients fall into a form of mutism, while others continue to communicate, albeit in an entirely disorganized and chaotic manner.

This brief study of confabulations seems to support our belief that the subject merits further attention from a purely linguistic perspective. As we indicated earlier, on the subject of patient speech as a whole, the creation of larger corpora, supported by detailed patient data (psychometric and biographical data, at the very least), is crucial for any further linguistic work. Our specific aim in this context would be to consider confabulations in their own right, almost as a form of narrative material for interpretation. The theme of textual genre, for example, offers promising scope for further exploration.

Conclusion

Improving Psychological Tests and Developing Discourse Analysis

In this work, we have examined the ways in which advances in textual semantics may be helpful in the development of psychological tests. Meaning is constructed from the global to the local level; words allow us to recognize signifiers, construct lexical units and create groupings, and not vice versa. In our view, a dynamic vision of language is crucial, whether for signifiers or signified elements, taking account of a constantly (but imperceptibly) changing body of pairings which is in a permanent state of deconstruction and reconstruction. Text (determined by a genre) forms the framework for recognition; the fact that existing tests generally ignore this fact, using individual words or expressions alone, is problematic. Given the importance of textual context, these single elements cannot constitute sufficient points of access to the lexical milieu. Any test designed to detect impairments in associative connections must therefore be altered and submitted to expert linguistic examination.

The argument, often put forward by neuroscientists to defend their methods, that the artificiality of communication situations used in testing is deliberate is not, in our opinion, entirely valid. Modeling a linguistic reality in light of specific aims does not give us the right to ignore this reality. If, as we believe, genre takes precedence over lower levels, then the results of psychological tests – a genre in their own right – should only be extrapolated to other genres with the greatest prudence. Furthermore, if the only linguistic production we consider is a set of substantives linked to simple visual representations, then the validity of our conclusions must necessarily be limited to a closely defined context.

Furthermore, the encoding – storage (with or without consolidation) – retrieval model (a perfect analogy of digital mechanisms) is rarely, if ever, questioned. In our view, this understanding of memory function needs to be re-examined in detail as far as language is concerned. We cannot therefore support the simplistic understanding of language expressed by Rossi (2013, p. 1):

> "In order to understand a situation, sentence or text, we need access to a mental dictionary containing associated meanings, and to objects, facts, states, situations and events".

Yet, as we have noted, the analogy between objects and lexemes cannot be considered valid. Language is so complex that even two successive occurrences of the same word in a sentence have different meanings: for example, in the case of "is that pasta salad, or *salad* salad?", the second "salad" – the first of the pair – indicates that the speaker is referring to a more prototypical, probably lettuce-based, dish. This phenomenon is known as contrastive focus reduplication. While this example is entirely banal, it clearly shows that the speaker or listener does not simply open the same "memory box" twice to understand the two occurrences. Where, then, is the second meaning of "salad" stored? Where do we find the rule indicating that the two iterations should not be understood in exactly the same way? What exactly is *recalled*? It comes as no surprise that word games, poetic constructions and the like – which form an essential aspect of human communication – are almost completely ignored in the cognitive sciences; they run contrary to a fixed vision of language, in which words are seen to have a clearly defined and generally unitary meaning. The complexity of language, a source of inexhaustible richness for the linguist, is reduced to simple nomenclature in neuroscience and, consequently, in psychological testing. At first glance, this position appears to create a simple and practical object for the purposes of testing, but this over-simplification comes at too high a cost: the results are as artificial as the situations in which they are obtained.

Despite the shortcomings we have identified in this book, we would still recommend the improvement, and not the abolition, of psychological tests. A multidisciplinary approach would be helpful in this respect. Such collaborative research would also serve in improving patient care. The development of tests that correspond more closely to real interpretative processes, featuring authentic communication conditions – i.e. using texts – would constitute a significant step in patient care. The establishment of these improved tests, and verification of their effectiveness in practice, remains crucial. If we succeed in understanding how and why Alzheimer's patients perform badly in well-constructed tests that take account of the textual aspect, the situation in which they are conducted and culture, we will likely gain a better understanding of how best to respond (at a non-pharmaceutical level) to patients' needs. There is also a connection between testing and medication: certain researchers test the effectiveness of drugs using the results of the tests described here, comparing patients' scores before and after administration. For example, Allain *et al.* (1997, p. 190) worked to promote mnesine, an "ideal" drug, in the early 2000s:

> "The early phases of development, using healthy volunteers (young and old) and sufferers in the very early stages of dementia, demonstrated a significant improvement in performance in all psychometric memory tests compared to the placebo, in a dose-dependent manner: the speed of information processing was objectively increased, as was the working memory".

At the other end of the spectrum, speech therapists have been working to improve support for patients, to maintain patient autonomy for as long as possible and (obliquely) to increase patient well-being. As Degiovanni (2014, p. 15) states: "This approach is based on cognitive stimulation, generally non-differentiated, using universal exercises; in the best cases, these may be adapted to correspond to the level and interests of the patients. The aim is to slow the advancement of the impairment through activation". The author aimed to activate several forms of memory concurrently:

> " – *recent episodic memory*, setting down real recent events and partial memories of certain lived experiences on paper, for example on a calendar;
>
> – *prospective memory*, creating a work schedule for the patient;
>
> – *memory 'of the past' (episodic then semantic):* creating a 'book of life' and a 'book of interests' with the patient in order to identify their areas of preoccupation closely as possible".

More often than not, this approach is adopted by practitioners for want of other tools, faced with the regular and inevitable progression of impairments.

It is, however, highly useful, notably in its focus on the "human" aspect, which patients themselves often consider most important, and which family members may struggle to maintain without support. Mutism – considered to be a symptom of the final stage of the disease –may, in fact, be the effect of a form of self-abandonment. Fifteen years ago, Ploton (2004, p. 137) was already calling for efforts to be made to combat this narcissic vulnerability: "[...] efforts to enable narcissic rebuilding should form the cornerstone of care". Any and all relational or recreational activities are likely to be beneficial in this area, as is the fact of giving appropriate responsibilities to patients.

Various activities have been developed in the area of cognitive/memory training, both for cognitively healthy individuals and for patients. The aim is to establish objective bases to produce real therapeutic benefits, drawing on cerebral plasticity as a preventative measure against decline in healthy patients, while mobilizing the cognitive reserve in sufferers, in an attempt to "repair" their capacities.

Neuropedagogy – or educational neuroscience – has emerged as a discipline in its own right in recent years. We have no preconceived notions concerning this area, although we are concerned that it may simply constitute an applied extension of the cognitive sciences, thus inheriting all of the usual, established postulates. De Rotrou and Frambourt (1997, p. 203) presented the subject as a series of "classic strategies, focusing on the operative aspects of cognitive function (association, classification, semantization, planning, analysis and synthesis)", all essentially based on imagery. The program is ambitious and extensive, and the inclusion of personal affective parameters is reassuring. Nevertheless, there is a risk that this new "neuropedagogy" could simply act as training for psychometric tests. Certain authors, such as Laurent and Thomas-Antérion (1997, p. 102), take a particularly dim view of cognitive training: "The potential accessibility of drugs or of memory training explains the interest of pharmaceutical companies and the proliferation of memory training groups, which often have more in common with jogging than with cognitive psychology". At present, there is no evidence to counter these criticisms.

In our view, the best option would be to focus heavily on individualized, "human" support, inspired by the practices established by Sioli, Nissl and Alzheimer himself at the Frankfurt asylum in the late 19th Century. Based on Kraepelin's recommendations, treatment involved balneotherapy, a diet of high-quality (and local) food and no alcohol (although drinking was not prohibited); crucially, the use of sedatives or sleep aids was limited to a strict minimum. Personal interviews were carried out daily. Evidently, the development of effective medication, alongside early detection methods, would be highly beneficial, and this is at the heart of current orientations (Piou 2012, p. 16):

> "Thus, the message emanating from certain lobbies highlights the medical aspect of a problem which is, in our own view, more of a social issue, although the current discourse tends to focus on families, to the detriment of patients".

Until we can be certain that the drug-based treatment route will provide a solution, other psychological tests would offer an alternative, or at least a good complement, to attempt to diagnose Alzheimer's disease at the physical level. Allain *et al.*'s affirmation (1997, p. 192) concerning support for mnesine strikes us as somewhat naive and inappropriate:

> "Oddly, work which demonstrates a beneficial effect on memory is rarer, or has not been publicized by any of the multiple streams of communication active in the world of pharmaceuticals".

Given the prevalence of Alzheimer's disease and media sensitivity to even the smallest discovery in the field, it seems highly unlikely that a medical breakthrough would have remained unpublicized, for whatever reason.

In our opinion, work on analyzing patient discourse should also focus on improving patient management: recurring linguistic features should be seen as symptoms to interpret. Care environments need to be adapted, to take account, for example, of patients' reduced vocabulary and simplified syntax; linguists have an important contribution to make in this regard. Similarly, the abundant use of neoforms – which are correctly formed from a morphological perspective – shows us that patients retain their morphosyntactic capacities. In our view, this information could be exploited to efficiently stimulate patients' linguistic and mnesic capacities. Nevertheless, it is important to remember that neither the objectives nor the scoring grids mentioned here are exhaustive; we have simply considered the language of Alzheimer's sufferers from a range of angles which, in the current state of research, appear relevant.

As with any area of fundamental research, we should not underestimate the effect of accidental discoveries – even the smallest – which can have a serious impact on the whole system. By highlighting particular linguistic features, we may indicate new pathways for research in neuroscience, and new findings in neuroscience may pave the way for innovative developments in linguistics. In general terms, strengthening the links between linguists and neuroscientists can only be beneficial, providing new inspirations in both domains. This type of collaboration could be promoted by various research foundations providing funding for humanities and social science projects, but for the moment, these organizations focus solely on domains with a more evident medical application. At the time of writing, linguists and other humanities/social science researchers must seek funding at the government and university level, or from external sources (public or private foundations), where the success rate is very low. In the first case, even if funding is obtained, there is a risk that results will not be widely disseminated; in the second case, the time lost applying for very limited resources and the lack of recognition for the work produced can be frustrating. It is always a challenge for researchers to push the boundaries of their own disciplines; theoretical "exploration" involves a significant and time-consuming learning curve, and any claim to expertise is dependent on this time being available. Cross-disciplinary adventures of this kind are therefore, once again, entirely dependent on funding being made available.

Our proposed approach is built around two main elements, both solidly rooted in collaborative working practices involving researchers and healthcare professionals. The first aim is to improve language-based psychological tests by increased attention to the linguistic aspect. The second is to support and contribute to a multidisciplinary approach to patient discourse analysis. We may thus envisage a new form of neurolinguistics – not simply an extension of classic linguistics, applied within the existing framework of neurology. We hope, with this book, to have established the broad outlines of the field, through a critical examination of the existing situation.

Appendices

Appendix 1

Interview Transcriptions: The Lion's Tale Test

The transcriptions below do not feature all of the annotation used elsewhere in this work, in order to improve readability. The original French text is shown in the left column, the English translation on the right.

All of these interviews were carried out by Marie Montillot, a final-year student at the speech therapy school in Besançon (France), who graciously agreed to provide us with the recordings. Individuals suspected of suffering from Alzheimer's disease were interviewed in their own homes. Any personal data (age, profession, etc.) which could be used to identify patients has been removed. To this end, we have chosen to focus on the Lion's Tale test alone, to the exclusion of data collected during anamnesis.

Looking at this collection of interviews, we observe a wide variety of different testing conditions: with or without an assistant or caregiver; at home (as in these cases) or elsewhere; and at different times of day (an important consideration for certain elderly subjects). We also note that tests featuring the Lion's Tale may be more "interactive" in nature than other tests: patients do not hesitate to interrupt the speech therapist to ask questions or give their opinion, even when she is reading the story. Finally, while some patients are cooperative and even seem to want to "succeed" in the test, others are less enthusiastic and pay little attention, with obvious implications for their results.

In any case, the material provided in these appendices is relatively rare: few practitioners take the time to transcribe their recordings (where recordings are available), and we hope that readers will derive as much benefit from the material as we ourselves did. In these extracts, "ST" is the speech therapist, "P" the patient and "A" the assistant or caregiver.

Patient 1

[Immediate recall]

(1) ST: xxx écouter une histoire qui est un petit peu longue et il faudra que vous retrouviez la trame générale de l'histoire, pas que vous répétiez mot à mot, mais que vous retrouviez un maximum de détails de l'histoire.	(1) ST: xxx listen to a story which is quite long, and I'll ask you to remember the general outline of the story, not repeat it word for word, but remember as many details as possible.
(2) P: Des phrases, des… des mots, des phrases ?	(2) P: Sentences, wo… words, sentences?
(3) ST: Bah, je vous raconte l'histoire.	(3) ST: Okay, I'll tell you the story.
(4) P: Oui ?	(4) P: Yes?
(5) ST: Et ensuite, vous, vous essayez de retrouver un maximum de détails, retrouver le sens de l'histoire.	(5) ST: And then you, you're going to try to give me as many details as possible, remember the outline of the story.
(6) P: Qui vont ? Qui vont pas ou… ?	(6) P: Which go? Which don't go, or…?
(7) ST: Bah, vous racontez l'histoire à votre tour.	(7) ST: Well, then it will be your turn to tell the story.
(8) P: ((toux)) Pardon. ((toux)) Eh bien ?	(8) P: ((coughs)) Excuse me. ((coughs)) Well then?
(9) ST: Oui ? D'accord.	(9) ST: Yes? Okay.
(10) P: Oui.	(10) P: Yes.
(11) ST: C'est parti ?	(11) ST: Shall we start?

(12) P: C'est obligé ?	(12) P: Do we have to?
(13) ST: Bah, j'suis là pour ça.	(13) ST: Well, that's what I'm here for.
(14) P: Non, non, on fait le contraire : je lis, puis vous vous xxx…	(14) P: No, no, let's do the opposite: I'll read, then you, you xxx…
(15) ST: ((rire)) Ce serait trop facile ! ((rire)) Bon, alors, c'est l'histoire du lion.	(15) ST: ((laughs)) That would be too easy! ((laughs)) Well, then, it's the story of a lion.
(16) P: Lion.	(16) P: Lion.
(17) ST: Un lion nommé Sultan s'échappa de sa cage par la porte mal fermée par un gardien négligent.	(17) ST: A lion named Sultan escaped from his cage through a door which was left open by a careless guard.
(18) P: Point.	(18) P: Period.
(19) ST: Point. La foule des visiteurs, nombreuse ce dimanche-là, s'enfuit vers les bâtiments voisins.	(19) ST: Period. The crowd of visitors who were numerous that Sunday fled to the neighboring buildings.
(20) P: La foule s'enfuit dans les bâtiments voisins, oui.	(20) P: The crowd fled to the neighboring buildings, yes.
(21) ST: Une femme, vêtue de bleu, qui tenait dans ses bras son enfant de un an, le laissa tomber. Le lion s'en saisit. La femme en larmes, revint sur ses pas, et supplia le lion de lui rendre son petit. L'animal la regarda longuement, fixement, et enfin lâcha l'enfant sans lui avoir fait le moindre mal.	(21) ST: A woman dressed in blue who was carrying her child, aged one, dropped him. The lion took him. The woman, in tears, turned back and begged the lion to give the child back. The animal looked at her for a long time, intently, and finally let go of the child, having done him no harm.
(22) P: Une belle histoire.	(22) P: It's a nice story.

(23) ST: Alors, dites-moi, qu'est-ce que vous pouvez euh… Essayez de retrouver la trame de l'histoire.	(23) ST: So, tell me, can you… try and remember the outline of the story.
(24) P: Une petite, une petite histoire merveilleuse. C'est un conte de fées, ça. C'est… La mère en bleu, c'est la fée. Une fée en bleu.	(24) P: A little, a marvelous little story. It's a fairy tale. It's… The mother in blue, she's the fairy. A fairy in blue.
(25) ST: Oui.	(25) ST: Yes.
(26) P: Ben c'est, c'est joli.	(26) P: Well it's, it's pretty.
(27) ST: Alors, de quoi ça parle ?	(27) ST: So, what's it about?
(28) P: Alors, c'est l'histoire d'une dame qui est dans, en visite dans un zoo et y'a une cage, vous avez oublié le mot cage, oui. Il y a une cage et une porte qui a mal été fermée par les employés.	(28) P: Well, it's the story of a lady who is in, visiting a zoo, and there's a cage, you forgot the word cage, yes. There's a cage and a door that wasn't closed properly by the employees.
(29) ST: Mmm.	(29) ST: Mmm.
(30) P: Une dame passe par là avec un enfant. Le lion s'en empare et moi, je pose la question, pour quelle raison, cette dame, elle est rentrée dans les cages, non ? Je sais pas. Qu'est-ce qui faut que je vous dise ?	(30) P: A lady goes by with her child. The lion takes him and me, I'm wondering why the lady went into the cages, no? I don't know. What do you want me to tell you?
(31) ST: Est-ce que vous vous souvenez d'autres choses… de l'histoire ?	(31) ST: Can you remember any other parts of the story?
(32) P: Qui… Sur l'histoire ? Ben qu'la porte était ouverte.	(32) P: That… About the story? Well, the door was open.
(33) ST: Oui.	(33) ST: Yes.
(34) P: Puis, les gardiens ont d'la chance de pas s'être fait mordre,	(34) P: Then, the keepers were lucky they didn't get bitten, torn

déchiqueter. Et c'est bizarre qu'il ait choisi le, le, le, le, le… gamin. C'est un gamin ? Je me rappelle plus c'que vous m'avez dit.	apart. And it's strange that it chose the, the, the, the… kid. It was a kid? I can't remember what you told me.
(35) ST: Je peux pas vous répondre, essayez de…	(35) ST: I can't tell you, try to…
(36) P: Oui, oui, oui. Parce que j'ai pas entendu, il l'a pris, puis il l'a pas, il l'a pas dévoré. Donc, c'est une belle histoire.	(36) P: Yes, yes, yes. Because I didn't here, he took it, then he didn't, he didn't eat it. So it's a nice story.
(37) ST: C'est une belle histoire.	(37) ST: It's a nice story.
(38) P: La fin.	(38) P: The end.
(39) ST: Oui, très bien.	(39) ST: Yes, very good.

[Delayed recall]

(1) ST: Parfait. Est-ce que vous vous souvenez de l'histoire que je vous ai racontée tout à l'heure ? Est-ce que vous pourriez retrouver… euh… la trame de l'histoire ?	(1) ST: Perfect. Do you remember the story I told you earlier? Can you remember, uh, the outline of the story?
(2) P: Mmmm. Non, mais je peux. On y va ?	(2) P: Mmmm. No, but I can. Shall we?
(3) ST: Dites-moi.	(3) ST: Tell me.
(4) P: Alors, c'est dans un… une cage. Un lion s'échappe. La porte étant laissée ouverte par la faute des employés. Arrive une dame avec un bébé.	(4) P: So, it's in a… a cage. A lion escapes. The door was left open, it was the employees' fault. A lady came with a baby.
(5) ST: Hm.	(5) ST: Hm.
(6) P: Vous avez de jolis yeux.	(6) P: You have pretty eyes.

(7) ST: ((rire))	(7) ST: ((laughs))
(8) P: Ça marche ?	(8) P: Okay?
(9) ST: Ne vous laissez pas distraire. Oui, oui, oui.	(9) ST: Don't let yourself get distracted. Yes, yes, yes.
(10) P: Et le lion s'empare du bébé.	(10) P: And the lion takes the baby.
(11) ST: Oui.	(11) ST: Yes.
(12) P: Arrive une dame habillée en bleu. Comme une fée. Le lion lâcha le bébé.	(12) P: A lady in blue arrives. Like a fairy. The lion let go of the baby.
((bruit – la porte s'ouvre, quelqu'un entre))	((noise – the door opens, someone comes in))
(13) P: Moralité : les animaux ont de la mémoire, parce que c'était une fée.	(13) P: Moral of the story: animals remember things, because it was a fairy.
(14) ST: Aaah !	(14) ST: Aaah!
(15) P: Parce que il peut pas le lâcher quand même… ((puis s'adressant à l'aidant qui vient d'entrer)) Oh ! Elle m'en fait voir la jeune dame.	(15) P: Because it can't let go of him all the same… ((then, talking to the assistant who has just come in)) Oh! This young lady is making me work hard.

Patient 2

(1) ST: À vous maintenant de raconter tout ce dont vous vous souvenez en un maximum de détails.	(1) ST: Your turn, now. Tell me what you remember with as much detail as possible.
(2) P: Euh… ((silence)) C'est l'histoire d'un lion…	(2) P: Uh… ((silence)) It's a story about a lion…
(3) ST: Oui.	(3) ST: Yes.

(4) P: Hein ?

(4) P: Eh?

(5) ST: Oui.

(5) ST: Yes.

(6) P: D'un lion qui ((souffle)) Ah, c'est déjà dur là. Hm…

(6) P: About a lion which ((blows out)) Ah, it's already getting difficult. Hmm…

(7) ST: Oui, c'est pas facile. Prenez votre temps, ne vous inquiétez pas.

(7) ST: Yes, it isn't easy. Take your time, don't worry.

(8) P: Un lion… qui s'échappa, c'est ça.

(8) P: A lion… which escaped, that's it.

(9) ST: Mmm. ((son d'approbation))

(9) ST: Mmm. ((sound of agreement))

(10) P: S'échappa de sa cage. ((silence)) Euh, il rencontra… une… une dame…

(10) P: It escaped from its cage. ((silence)) Uh, it met a, a woman…

(11) ST: Oui.

(11) ST: Yes.

(12) P: … qui tient un bébé dans ses bras.

(12) P: … who was holding a baby in her arms.

(13) ST: Mmm.

(13) ST: Mmm.

(14) P: Elle le laissa tomber. ((silence)) Euh, le lion s'approcha de du bébé. Il lui… pique le bébé, enfin, reste auprès du bébé. Et la m… la la maman se se… appela du monde pour pour l'aider à récupérer son bébé.

(14) P: She dropped him. ((silence)) Uh, the lion came up to the baby. It… stole the baby, well, stayed near the baby. And the m… the mother, she… called people to help her get her baby back.

(15) ST: Oui.

(15) ST: Yes.

(16) P: Et puis, lui, le lion, il est, au lieu de lui faire du mal, ben il, il, il la regarda bien dans les yeux.

(16) P: Then he, the lion, he's, instead of hurting him, well, he, he, he looked straight into her eyes.

(17) ST: Oui.

(17) ST: Yes.

(18) P: Et puis, il lâcha le bébé et puis, la, la maman reprit son bébé.	(18) P: And then he let go of the baby and then the, the mother took her baby.
(19) ST: ((silence)) Ok. Parfait.	(19) ST: ((silence)) Okay. Perfect.
(20) P: Je croyais que j'allais oublier… oui.	(20) P: I thought I was going to forget… yes.
(21) ST: Oui.	(21) ST: Yes.
(22) P: Pas mal de choses.	(22) P: Quite a lot of things.
(23) ST: Ben non.	(23) ST: Well no.
(24) P: Non ((ton étonné)). Ben c'est bien.	(24) P: No ((surprised tone)). Well that's good.

Patient 3

[Immediate recall]

(1) ST: Des papiers. C'est pas le plus intéressant. Alors. Eh ben, alors, c'est parti. Donc, je vais vous lire une histoire qui est assez longue. Le but, c'est, ce n'est pas de répéter mot à mot cette histoire, mais de vous souvenir d'un maximum de détails.	(1) ST: Papers. Nothing very interesting. So, well, here we go. So I'm going to read you a story, it's quite long. The aim isn't to repeat it word for word, but for you to remember as many details as possible.
(2) P: D'accord.	(2) P: Okay.
(3) ST: Et de redonner la trame générale de l'histoire ensuite. D'accord ?	(3) ST: And then to give me a general outline of the story afterward. Okay?
(4) P: On va essayer.	(4) P: We'll try.
(5) ST: Alors, il s'agit de l'histoire du lion. Un lion nommé Sultan	(5) ST: So, it's a story about a lion. A lion named Sultan escaped from

s'échappa de sa cage par la porte mal fermée par un gardien négligent.	his cage through a door which was left open by a careless guard.
(6) P: Donc il était au zoo ? On sait pas. D'accord. C'est un gardien, donc il était en cage, ok.	(6) P: So he wasn't in the zoo? We don't know. Okay. There's a guard, so he was in a cage, okay.
(7) ST: La foule des visiteurs, nombreuse ce dimanche-là, s'enfuit dans les bâtiments voisins. Une femme, vêtue de bleu, qui tenait dans ses bras son enfant de un an, le laissa tomber. Le lion s'en saisit. La femme en larmes, revint sur ses pas, et supplia le lion de lui rendre son petit. L'animal la regarda longuement, fixement, et enfin lâcha l'enfant sans lui avoir fait le moindre mal.	(7) ST: The crowd of visitors who were numerous that Sunday fled to the neighboring buildings. A woman dressed in blue who was carrying her child, aged one, dropped him. The lion took him. The woman, in tears, turned back and begged the lion to give the child back. The animal looked at her for a long time, intently, and finally let go of the child, having done him no harm.
(8) P: Voici.	(8) P: Here.
(9) ST: Qu'est-ce que vous pouvez, est-ce que vous pouvez redonner la trame de l'histoire en un maximum de détails ?	(9) ST: What can you, can you give me the outline of the story in as much detail as possible?
(10) P: Oh, déjà, on sait qu'on est dans un zoo… que, apparemment, c'était mal surveillé ou mal fermé. Le lion s'est sauvé.	(10) P: Well, we know we're in a zoo… and apparently it was badly guarded or badly closed. The lion got out.
(11) ST: Oui.	(11) ST: Yes.
(12) P: Euh, il y avait une dame qui, qui était avec son enfant et le lion l'a saisi.	(12) P: Uh, there was a lady who, who was there with her child and the lion took it.
(13) ST: Oui.	(13) ST: Yes.
(14) P: La dame l'a supplié de rendre son enfant et il lui a rendu.	(14) P: The lady begged it to give her child back and it gave him

C'est ce que j'ai retenu de principal.

back. That's the main thing I remembered.

(15) ST: Très bien.

(15) ST: Very good.

(16) P: C'est ça ?

(16) P: Is that it?

(17) ST: Mais oui, c'est très bien. On a fait… Oui, oui, très, très bien. On a la trame de l'histoire. Parfait. Je vais vous poser des, des petites questions maintenant pour vous connaître un petit peu mieux, parce que je viens comme ça mais, quand même…

(17) ST: Yes, that's very good. We've… yes, very, very good. We have the outline of the story. Perfect. I'm going to ask you some, some short questions now to get to know you a little better, because I'm just coming in like that, but still…

[Delayed recall]

(1) ST: Très bien, super. Est-ce que vous vous souvenez de l'histoire que je vous ai lue tout à l'heure ?

(1) ST: Very good, great. Do you remember the story I read to you earlier?

(2) P: Avec le lion ?

(2) P: About the lion?

(3) ST: Avec le lion.

(3) ST: About the lion.

(4) P: Oui ! Oui, je me rappelle.

(4) P: Yes! Yes, I remember.

(5) ST: Est-ce que vous pourriez redonner quelques… est-ce que vous pouvez redire l'histoire ?

(5) ST: Can you give me a few… can you tell me the story?

(6) P: L'histoire c'est que ce lion eh… sa cage devait être mal fermée, il s'est sauvé.

(6) P: The story is that this lion uh… its cage must have been closed badly, he got out.

(7) ST: Oui.

(7) ST: Yes.

(8) P: Il a envie de se promener. Et il a capturé un petit enfant. Peut-être qu'il avait l'intention de le

(8) P: He wanted to go walkabout. And he caught a little child. Maybe he planned to eat him, I don't

manger, je sais pas. Enfin, il l'a capturé et la maman désespérée lui a demandé de lui rendre et il lui a rendu. C'est ce que j'ai retenu.

know. Well, he caught it, and the desperate mother asked him to give him back, and he gave him back. That's what I remember.

(9) ST: Oui, très bien. Mais super.

(9) ST: Yes, very good. Great.

((rires))

((laughs))

(10) P: Et bah, nous avons terminé. Pour moi.

(10) P: Well then, we've finished. For me.

(11) P: Ah bon ! Bah, c'est bien !

(11) P: Oh, right! Well, that's great!

Patient 4

[Immediate recall]

(1) ST: Voilà. Je vous la raconte, cette histoire ? C'est l'histoire d'un lion.

(1) ST: Okay. Shall I tell you the story? It's the story of a lion.

(2) P: L'histoire d'un lion ?

(2) P: The story of a lion?

(3) ST: Oui.

(3) ST: Yes.

(4) P: Oh, là, là.

(4) P: Goodness me.

(5) ST: « Oh, là, là », comme vous dites. Est-ce que vous voulez vous asseoir ?

(5) ST: "Goodness me", that's right. Do you want to take a seat?

(6) P: xxxx

(6) P: xxxx

(7) ST: Alors, c'est l'histoire du lion. ((fort bruit)) xxx retrouviez la trame de l'histoire, que vous redonniez un maximum de détails, une fois que je l'aurais lue. C'est un petit peu long, alors écoutez bien et essayez de vous souvenir d'un maximum de choses. D'accord ?

(7) ST: So, it's the story of a lion. (Loud noise)) xxx pick up the thread of the story, you give me as many details as you can, after I read it. It's a little long, so listen well and try to remember as much as possible. Okay?

(8) P: Hm.	(8) P: Hmm.
(9) ST: Alors, c'est l'histoire du lion. Un lion nommé Sultan s'échappa de sa cage par la porte mal fermée…	(9) ST: So, it's a story about a lion. A lion named Sultan escaped from his cage through a door which was left open…
(10) P: Je ne l'ai pas vu passer, là.	(10) P: I haven't seen it.
(11) ST: Bah, écoutez bien… par la porte mal fermée par un gardien négligent. La foule des visiteurs, nombreuse ce dimanche-là, s'enfuit dans les bâtiments voisins. Une femme, vêtue de bleu, qui tenait dans ses bras son enfant de un an, le laissa tomber. Le lion s'en saisit. La femme en larmes, revint sur ses pas, et supplia le lion de lui rendre son petit. L'animal la regarda longuement, fixement, et enfin lâcha l'enfant sans lui avoir fait le moindre mal. Voilà.	(11) ST: Well, listen… through a door which was left open by a careless guard. The crowd of visitors who were numerous that Sunday fled to the neighboring buildings. A woman dressed in blue who was carrying her child, aged one, dropped him. The lion took him. The woman, in tears, turned back and begged the lion to give the child back. The animal looked at her for a long time, intently, and finally let go of the child, having done him no harm. That's it.
(12) P: Oui.	(12) P: Yes.
(13) ST: Alors, qu'est-ce que vous pouvez me redire de cette histoire ? Essayez de donner un maximum de détails. ((bruit)) Parfait, je vous remercie. Merci. De quoi ça parle ?	(13) ST: So, what can you tell me about this story? Try to give me as many details as you can. ((noise)) Perfect, thank you. Thanks. What's it about?
(14) P: De quoi ça parle ? Si… si, pourquoi… il y avait aussi un autre… une fille… oui, il y avait une fille… une fille, amie, il y en a deux… à sa place celle-là, de l'autre côté là-bas.	(14) P: What's it about? Yes…yes, why… there was another, too… a girl… yes, there was a girl… a girl, friend, there are two of them… in her place, that one, on the other side over there.
(15) ST: D'accord.	(15) ST: Okay.

(16) P: Oui ?	(16) P: Yes?
(17) ST: Alors, ça, vous m'aviez parlé de votre fille tout à l'heure. Oui.	(17) ST: So, that, you told me about your daughter earlier. Yes.
(18) P: Hm.	(18) P: Hmm.
(19) ST: Et là, à propos de l'histoire que je viens de vous lire ? Est-ce que vous vous souvenez de, de détails ?	(19) ST: Now, about the story I just read you? Do you remember any, any details?
(20) P: Un détail ? Non, j'ai pas de détails.	(20) P: A detail? No, I don't have any details.
(21) ST: Est-ce que vous vous souvenez d'une petite chose ?	(21) ST: Do you remember one little thing?
(22) P: … pas de temps.	(22) P: … no time.
(23) ST: De quoi ça parlait ? Oooh, pas le temps. Je comprends. ((rire)). De quoi elle parlait, cette histoire que je vous ai lue ?	(23) ST: What was it about? Ooooh, no time. I understand. ((laughs)). What was it about, the story I read you?
(24) P: Non, j'ai pas, j'ai pas lu ça. Hm. Y'a pas beaucoup, alors… Je n'ai pas dit une épouse.	(24) P: No, I haven't, I haven't read that. Hmm. There isn't much, so… I didn't say a wife.
(25) ST: D'accord. Bien. Est-ce que vous rappelez qu'il y avait un animal, dans cette histoire ?	(25) ST: Okay. Well. Do you remember that there was an animal, in the story?
(26) P: Un animal ?	(26) P: An animal?
(27) ST: Oui.	(27) ST: Yes.
(28) P: C'est vrai ?	(28) P: Really?
(29) ST: Oui. Dans l'histoire que je vous ai lue, ça parlait d'un animal.	(29) ST: Yes. In the story I read to you, there was an animal.
(30) P: Hm. Un animal…	(30) P: Hmm. An animal…

(31) ST: Hm. Est-ce que vous vous souvenez ?	(31) ST: Hmm. Do you remember?
(32) P: Eh, oui.	(32) P: Eh, yes.
(33) ST: C'était quoi ?	(33) ST: What was it?
(34) P: C'était quoi ? Ha… C'était, quand on est parti avec là, on est allé jusqu'à l'autre côté, là.	(34) P: What was it? Ah… It was, when we went with the, we went over to the other side, there.

[Delayed recall]

(1) ST: Est-ce que vous vous rappelez, tout à l'heure, je vous ai raconté une histoire ? Est-ce que vous vous souvenez de quelque chose ?	(1) ST: Do you remember, earlier, I told you a story? Do you remember anything?
(2) P: Comme tout le monde.	(2) P: Like everyone.
(3) ST: Alors, dites-moi.	(3) ST: So tell me.
(4) P: Pas du tout.	(4) P: Not at all.
(5) ST: De quoi est-ce que vous vous souvenez ? Même une petite chose.	(5) ST: What do you remember? Even one little thing.
(6) P: Hm, petite…	(6) P: Hmm, little…
(7) ST: De quoi parlait cette histoire ?	(7) ST: What was the story about?
(8) P: J'sais pas.	(8) P: Don't know.
(9) ST: C'est oublié ? C'était l'histoire d'un animal.	(9) ST: Did you forget? It was the story of an animal.
(10) P: Oui.	(10) P: Yes.
(11) ST: Oui. Un animal dans un zoo, vous m'avez dit.	(11) ST: Yes. An animal in a zoo, you told me.

(12) P: Oui ? xxxx	(12) P: Yes? xxxx
(13) ST: Oui. Quel était cet animal?	(13) ST: Yes. What was the animal?
(14) P: Cet animal ?	(14) P: The animal?
(15) ST: C'était quoi ? Oui. C'est un animal de la savane.	(15) ST: What was it? Yes. An animal from the savannah.
(16) P: De la savane ?	(16) P: From the savannah?
(17) ST: Hm.	(17) ST: Hmm.
(18) A: Hm. Surtout que t'en as vu.	(18) A: Hmm. And you've seen some.
(19) ST: Aaah !	(19) ST: Aaah !
(20) A: On a été six ans. Où on est allé pendant six ans ?	(20) A: We were there for six years. Where did we go for six years?
(21) ST: Oui.	(21) ST: Yes.
(22) A: C'est bien, ça. Parce qu'à table eh bien il en parle toujours, de cet événement. Au Sénégal !	(22) A: That's good, that one. Because at mealtimes, well, he talks about it all the time, that time. In Senegal!
(23) ST: Oh, là, là ! Au Sénégal !	(23) ST: Goodness me! In Senegal!
(24) A: Eh, bon, au Sénégal, oui.	(24) A: Well, in Senegal, yes.

Patient 5

[Immediate recall]

(1) ST: Donc, je vous redis mais non, je vous explique. Ça veut une… Je vais lire une histoire qui est assez longue. Il faudra que vous vous souveniez d'un maximum de détails et redonniez la trame générale de l'histoire.	(1) ST: So, I'll tell you again but no, I'll explain. It wants… I'm going to read quite a long story. You'll need to remember as many details as you can and give me the general thread of the story.

(2) P: Oui.	(2) P: Yes.
(3) ST: Voilà. Vous faites de votre mieux et puis après vous me redirez ce dont vous vous souvenez. Vous êtes prête ?	(3) ST: There. Do your best, then after, you'll tell me what you remember. Are you ready?
(4) P: Hm.	(4) P: Hmm.
(5) ST: Il s'agit de l'histoire du lion.	(5) ST: It's the story of a lion.
(6) P: Du lion ?	(6) P: A lion?
(7) ST: Oui. Un lion, nommé Sultan…	(7) ST: Yes. A lion, named Sultan…
(8) P: Un lion ?	(8) P: A lion?
(9) ST: Nommé Sultan.	(9) ST: Named Sultan.
(10) P: Nommé Sultan.	(10) P: Named Sultan.
(11) ST: … s'échappa de sa cage par la porte mal fermée par un gardien négligent. La foule des visiteurs, nombreuse ce dimanche-là, s'enfuit dans les bâtiments voisins. Une femme, vêtue de bleu, qui tenait dans ses bras son enfant de un an, le laissa tomber. Le lion s'en saisit. La femme en larmes, revint sur ses pas, et supplia le lion de lui rendre son petit. L'animal la regarda longuement, fixement, et enfin lâcha l'enfant sans lui avoir fait le moindre mal.	(11) ST: … escaped from his cage through a door which was left open by a careless guard. The crowd of visitors who were numerous that Sunday fled to the neighboring buildings. A woman dressed in blue who was carrying her child, aged one, dropped him. The lion took him. The woman, in tears, turned back and begged the lion to give the child back. The animal looked at her for a long time, intently, and finally let go of the child, having done him no harm.
(12) P: Hm.	(12) P: Hmm.

(13) ST: Qu'est-ce que vous pouvez me redire de cette histoire ?	(13) ST: What can you tell me about this story?
(14) P: Je peux tout vous raconter l'histoire mais le nom du cheval ((rire)) le nom du lion. Il s'appelle… C'est l'histoire d'un lion.	(14) P: I can tell you all of the story, but the name of the horse ((laughs)) the name of the lion. He was called… It's the story of a lion.
(15) ST: Oui ? Tout à fait.	(15) ST: Yes? Exactly.
(16) P: Qui… qui se retrouve… L'histoire d'un lion qui se retrouve dans le zoo. Au zoo, non ?	(16) P: Who… who was… The story of a lion who was in the zoo. In a zoo, no?
(17) ST: Hm.	(17) ST: Hmm.
(18) P: Il s'appelle pas Sascha, non ? Quelque chose comme ça, non ?	(18) P: He was called Sascha, no? Something like that, no?
(19) ST: Non, pas tout à fait. C'est pas grave si vous vous souvenez pas.	(19) ST: No, not exactly. It doesn't matter if you don't remember.
(20) P: Alors, là, y'a une maman avec un bébé qui arrive et elle lâche le bébé. Le chien, le ch… le lion le prend sans lui faire de mal et le rend à sa maman, non ? C'est ça ?	(20) P: So then, there's a mother with a baby who comes and she lets go of the baby. The dog, the do… the lion takes him without hurting him and gives him back to his mother, no? Is that it?
(21) ST: Bah, oui.	(21) ST: Well, yes.
(22) P: J'sais pas comment il s'appelle, le lion.	(22) P: I don't know what he was called, the lion.
(23) A: Ça commence par S.	(23) A: It starts with an S.
(24) P: Par A ?	(24) P: An A?
(25) A: Par S. T'avais dit « Sascha », c'était la première lettre et…	(25) A: An S. You said "Sascha", that was the first letter and…

(26) ST: Vous y étiez presque…	(26) ST: You were almost there…
(27) P: Ça finit par Sascha, oui, ça me dit quelque chose. Sascha, ça…. Ça finit pas par un A ?	(27) P: It ends in Sascha, yes, that rings a bell. Sascha, that… It doesn't end in A?
(28) ST: Non. On peut laisser et puis on verra si tout à on y revient. Si vous voulez.	(28) ST: No. We can leave it and then we'll see if, later, we're back. If you like.
(29) P: xxx En fait, bon.	(29) P: xxx Actually, well.
(30) ST: Sait-on jamais…	(30) ST: You never know…
(31) P: Oui.	(31) P: Yes.
((rire))	((laughs))
(32) P: Alors, il le lâche, le bébé, quand même ?	(32) P: So he did let go of the baby, though?
(33) ST: Oui, vous avez raison, oui. C'est ça. Très bien. Bon. Vous avez dit tout ce dont vous souvenez ? On a fait le tour, vous pensez ?	(33) ST: Yes, you're right, yes. Very good. Well. Have you told me everything you remember? Does that cover it, do you think?

Patient 6

[Immediate recall]

(1) ST: C'est ça. Un lion, nommé Sultan… Écoutez bien toute l'histoire et puis après…	(1) ST: That's right. A lion, named Sultan… Listen to the whole story, and then…
(2) P: Oui, oui.	(2) P: Yes, yes.
(3) ST: Voilà. Un lion nommé Sultan s'échappa de sa cage par la porte mal fermée par un gardien négligent. La foule des visiteurs, nombreuse ce dimanche-là, s'enfuit	(3) ST: Here. A lion named Sultan escaped from his cage through a door which was left open by a careless guard. The crowd of visitors who were numerous that

dans les bâtiments voisins. Une femme, vêtue de bleu, qui tenait dans ses bras son enfant de un an, le laissa tomber. Le lion s'en saisit. La femme en larmes, revint sur ses pas, et supplia le lion de lui rendre son petit. L'animal la regarda longuement, fixement, et enfin lâcha l'enfant sans lui avoir fait le moindre mal.	Sunday fled to the neighboring buildings. A woman dressed in blue who was carrying her child, aged one, dropped him. The lion took him. The woman, in tears, turned back and begged the lion to give the child back. The animal looked at her for a long time, intently, and finally let go of the child, having done him no harm.
(4) A: … une histoire.	(4) A: … a story.
(5) P: C'est long, quand même.	(5) P: It's long, though.
(6) A: … petite histoire.	(6) A: … little story.
(7) ST: C'est long, c'est long. Alors, essayez de me redonner un maximum de détails sur…	(7) ST: It's long, it's long. So, try to tell me as many details as you can about…
(8) P:… c'est l'histoire d'un, d'un… Comment il s'appelle… Sultan ?	(8) P:… it's the story of a, of a… What was he called… Sultan?
(9) ST: Oui ?	(9) ST: Yes?
(10) P: Qui est un lion.	(10) P: Who is a lion.
(11) ST: Oui ?	(11) ST: Yes?
(12) P: Et pas… et les… et… Il rencontre des gens qu'ont perdu leur enfant. Qui… qu'ont lâché leur enfant.	(12) P: And not… and the… and… he meets people who have lost their child. Who… who let go of their child.
(13) ST: Oui ?	(13) ST: Yes?
(14) P: Donc… ils ont peur parce que c'est un lion, c'est Sultan le lion.	(14) P: So… they were afraid because he's a lion, he's Sultan the lion.
(15) ST: Hm ?	(15) ST: Hmm?
(16) P: Donc… Et puis, après ils	(16) P: So… And then, after they're

font très très attention et Sultan leur rend leur enfant, sans mal, sans rien. En gros.	very very careful and Sultan gives them their child back, without hurting him or anything. More or less.
(17) ST: C'est ça.	(17) ST: That's right.
(18) P: C'est ça ?	(18) P: That's right?
(19) ST: Tout à fait. Est-ce que vous pensez avoir à peu près tout dit ? Tout ce dont vous souvenez ?	(19) ST: Exactly. Do you think you've said pretty much everything?
(20) P: Une bonne partie. Mais un peu de…	(20) P: A fair part. But a little…
(21) A: Peut-être en moins de détails.	(21) A: Maybe in less detail.
(22) P: … en moins de détails.	(22) P: … in less detail.

[Delayed recall]

(1) ST: Merci. Parfait, ça ! Super ! Est-ce que vous vous souvenez de quelques éléments de l'histoire que je vous ai racontée tout à l'heure ?	(1) ST: Thanks. That's perfect! Great! Do you remember some parts of the story I told you earlier?
(2) P: Oui.	(2) P: Yes.
(3) ST: Qu'est-ce que vous pourriez me dire ?	(3) ST: What can you tell me?
(4) P: Une histoire d'un… d'un lion.	(4) P: A story about a… a lion.
(5) ST: Oui ?	(5) ST: Yes?
(6) P: Qui est gentil mais il rencontre une, une famille ou un petit garçon. Et puis le petit garçon… le lion le voit, prend le petit garçon. La famille est inquiète parce que, bon, on voit un enfant dans… un enfant dans, dans	(6) P: Who is kind, but he meets a, a family or a little boy. And then the little boy… the lion sees him, takes the little boy. The family is worried because, well, you see a child in, in the

les bras du lion. Mais le lion est très gentil et il redonne les enfants.

(7) ST: Très bien. Bon, bah, super. Ça va ? Je vous ai pas trop épuisé ?

(8) P: Non.

(9) ST: Bon ! Eh bien, je vous ai demandé tout ce que j'avais vous demandé, donc, je vous remercie beaucoup.

arms of a lion. But the lion is very kind and he gives the children back.

(7) ST: Very good. Well, then, great. How are you? I haven't tired you too much?

(8) P: No.

(9) ST: Good! Well then, I've asked you everything I had to ask you, so thank you very much.

Appendix 2

English Translations for Chapter 4

This Appendix contains English translations of the extracts from *Puzzle, journal d'une Alzheimer* by Claude Couturier used in Chapter 4. The original French text is provided on pages 96–99 and 102–104.

Extract from pages 96–99

/imprecision/
(1) *One day* (?) in summer 1996

/iterative/ *confusion/*
(2) *Once again*, today, *I got myself lost* in the parking lot at

(3) Carrefour [supermarket]; I was with Céline and the day had already gotten off to a bad

(4) start

/privation/
(5) I can only remember *bits of it*, as if the day happened

/privation/
(6) in *dots …*

/confusion/
(7) I *realized* we were on the road ten

(8) kilometers in (even though I was driving) when, before a crossroads, I

(9) put my blinkers on to turn right! Céline said:

(10) – Where are you going? We're not going to the village, but straight into town!

/privation/

(11) I tried to remember leaving the house, but *nothing*, just a

/privation/ /imprecision/

(12) *black hole*! I don't remember driving for *about*

(13) ten minutes. I find that worrying, but Céline tells me I'd been driving

(14) normally up to that point.

/imprecision/

(15) Then, apparently, things went *more or less* okay, but I

/confusion/ */imprecision/*

(16) just *can't* say what we did *exactly*: all I

/imprecision/ */confusion/*

(17) know is that *at some point*, I *got the impression* I was having

/confusion/

(18) a *dizzy spell* in the supermarket aisles and in the shopping

(19) mall where we both like to go for

/confusion/

(20) a quiet stroll. I said to Céline that *I must be tired*

(21) and that we had better go home …

/confusion/

(22) But when we left, I *couldn't find* the keys to the truck in my

/confusion/

(23) purse; I *searched* all my jacket pockets, nothing.

(24) Céline looked at me, laughing, and she said:

(25) – Yes, you must be tired … the keys are in your hand.

/confusion/

(26) – Of course, but *I can't see* the truck; where did I

(27) park it? – It's there, look, just opposite … next to the

(28) caddy park. What's the matter?

/confusion/ *lconfusionl*

(29) I *only saw* the truck once we'd got to it. There, really, *I don't*

/confusion/

(30) *know* what was happening, it's like *I didn't recognize*

(31) the thing I was looking for.

(32) When we got back, Céline was worried, she wouldn't stop asking me

(33) if I was okay. She said I should rest a while, that

lconfusionl *literativel*

(34) tiredness was making me *lose my mind for a moment*. I'll admit that

/confusion/

(35) I *struggle* to remember her timetable, but also

(36) I think she keeps changing what we planned. And,

(37) I have the same problem with Jo who takes orders for one day and

(38) decides all of a sudden to deliver a different day than the one

(39) that was planned. If they all start doing that, it isn't surprising

/confusion/

(40) that I get a little *mixed up*.

Extract from pages 102–104

/progressive/ /durational/ /claustration/ /confusion/
(1) "See, I *keep coming back* to a *labyrinth* between *dreams* and

/danger/
(2) reality, with a *bull* in the corner to top it all off, it's looking *more*
/progressive/

/danger/ /failure/
(3) *and more* like the Greek legend of the *Minotaur*! *If only* I

/claustration/
(4) could have a *thread*, like Theseus, so I could

/claustration/x /struggle/ /claustration/
(5) *find the way out* of this *nightmare*.

/danger/
(6) I *fear* the day when, even re-reading these lines, they *won't*

/failure/ /struggle/
(7) *remind me of anything*, because I sometimes *have to* think

/struggle/ x /failure/
(8) for a long time to *try to remember* certain events

(9) that I've written about here … on the other hand, I remember that I *forgot*
/failure/

(10) to write about certain things several days after they

/claustration/
(11) happened … I'm living with a lag. What a *maze*! And now I'm *back*
/iterative/

/claustration/
(12) inside my *labyrinth.* If I'm the architect, then my son Icarus should

/danger/
(13) come quickly to *save me* with his wings, before the monster

(14) devours me… I shouldn't have read all of those legends when I was

(15) young: it's giving me an overactive imagination and fueling my

/danger/
(16) *anxiety.*

(17) But I'm not the only one to feel like I'm in a

/claustration/
(18) *labyrinth*: an American patient wrote a book that I've just

(19) borrowed from the library, it's her story; like me,

(20) she needed to write about what she was experiencing, because everything was *escaping* her
/failure/

/progressive/ /isolation/
(21) *little by little*, but *she* had support and encouragement from

(22) the people around her when she was writing.

/isolation/ * /inevitable/*
(23) *Personally*, I'm doing it a bit out of *desperation;* it's

/isolation/ */isolation/* */isolation /*
(24) *my* trail of breadcrumbs, *my* expansion tank, *my* safety

/isolation/ */isolation/*
(25) valve, the *only* connection that's *intimate* enough to tie me back to *my* experience
/isolation/

/isolation/
(26) (*no-one* could tell me about it the way I saw it once

/inevitable/ */failure/*
(27) *I've* forgotten it and what's more, I *can't* express myself

/failure/ */iterative/*
(28) aloud the way I'd like. It *jams from time to time*; I'm

/progressive/

(29) getting *more and more* emotional about the subject, and so, since if you want something done properly,

/isolation/ */isolation/*

(30) you should do it *yourself, I'm telling myself my* story.

References

Allain, H., Bellissant, E., Bentue-Ferrer, D., Reymann, J.-M., Zekri, O., and Lieury, A. (1997). Pharmacologie de la plainte mnésique. In *De la plainte mnésique à la maladie d'Alzheimer*, Michel, B.F., Dérouesné, C., and Gély-Nargeot, M.C. (eds). Solal, Marseille, 189–196.

Anastasi, A. (1988). *Psychological Testing*, 6th edition. Macmillan Publishing Company, New York.

Andree, B., Hittmair, M., and Benke, T. (1992). Erkennen und Erklären von Sprichwörtern bei Patienten mit Alzheimer Demenz. *Neurolinguistik – Zeitschrift für Aphasieforschung und therapie*, 6, 27–34.

Asp, E.D. and de Villiers, J. (2010). *When Language Breaks Down – Analysing Discourse in Clinical Contexts*. Cambridge University Press, Cambridge.

Balthazar, M.L., Martinelli, J.E., Cendes, F., and Damasceno, B.P. (2007). Lexical semantic memory in amnestic mild cognitive impairment and mild Alzheimer's disease. *Arquivos de Neuro-psiquiatria*, 65(3A), 619–622 [Online]. Available: http://www.scielo.br/pdf/anp/v65n3a/14.pdf.

Barbizet, J. and Truscelli, D. (1965). L'Histoire du Lion (considérations sur la fabulation). *Semaine des hôpitaux*, 8, 1688–1694.

Barkat-Defradas, M., Martin, S., Rico-Duarte, L., and Brouillet, D. (2008). Les troubles du langage dans la maladie d'Alzheimer. *27èmes journées d'études sur la Parole*. Avignon [Online]. Available: https://hal.archives-ouvertes.fr/hal-00321233.

Bear, M.F., Connors, B.W., and Paradiso, M.A. (2007). *Neurosciences : à la découverte du cerveau*. Groupe Liaison, Rueil-Malmaison.

Begley, C.G. (2013). Reproducibility: Six red flags for suspect work. *Nature*, 497, 433–434.

Begley, C.G. and Ellis, L.M. (2012). Raise standards for preclinical cancer research. *Nature*, 483, 531–533.

Begley, C.G. and Ioannidis, J.P.A. (2015). Reproducibility in science: Improving the standard for basic and preclinical research. *Circulation Research*, 116, 116–126.

Begley, C.G., Buchan, A.M., and Dimagl, U. (2015). Institutions must do their part for reproducibility. *Nature*, 525, 25–27.

Bernaud, J.-L. (2000). *Tests et théories de l'intelligence*. Dunod, Paris.

Brouillet, D. and Syssau, A. (2005). *La maladie d'Alzheimer*. PUF, Paris.

Cantos Gómez, P. (2010). Analysing linguistic decline in early stage Alzheimer's disease: A corpus-based approach. In *A Mosaic of Corpus Linguistics: Selected Approaches*, Moisés, A. and Sánchez, A. (eds). Peter Lang, Bern.

Castera, M., Kuhn, V., and Medina, F. (2010). Mobilisation de l'accès lexical en orthophonie auprès de patients Alzheimer [Online]. Available: http://www.gnosia.fr/download/poster%20nice%20version%20finale.pdf.

Ciaramelli, E. and Ghetti, S. (2007). What are confabulators' memories made of? A study of subjective and objective measures of recollection in confabulation. *Neuropsychologia*, 45, 1489–1500.

Ciaramelli, E., Ghetti, S., Frattarelli, M., and Ladavas, E. (2006). When true memory availability promotes false memory: Evidence from confabulating patients. *Neuropsychologia*, 44, 1866–1877.

Ciaramelli, E., Ghetti, S., and Borsotti, M. (2009). Divided attention during retrieval suppresses false recognition in confabulation. *Cortex*, 45, 141–153.

Cieślicka, A. (2006). Literal salience in on-line processing of idiomatic expressions by second language learners. *Second Language Research: SAGE Publications*, 22(2), 115–144 [Online]. Available: https://hal.archives-ouvertes.fr/hal-00572096/document.

Code, C. and Lodge, B. (1987). Language in dementia of recent referral. *Age and Ageing*, 16, 366–372.

Cole, S.N., Fotopoulou, A., Oddy, M., and Moulin, C.J.A. (2014). Implausible future events in a confabulating patient with an anterior communicating artery aneurysm. *Neurocase*, 20(2), 208–224.

Conway, M.A. and Tacci, P.C. (1996). Motivated confabulation. *Neurocase*, 2, 325–339.

Cooper, J.M., Shanks, M.F., and Venneri, A. (2006). Provoked confabulations in Alzheimer's disease. *Neuropsychologia*, 44, 1697–1707.

Couturier, C. (2004). *Puzzle, journal d'une Alzheimer*, 3rd edition. Josette Lyon, Paris.

Crisp, J. (1995). Making sense of the stories that people with Alzheimer's tell: A journey with my mother. *Nursing Inquiry*, 2, 133–140.

Cunningham, J.M., Pliskin, N.H., Cassisi, J.E., Tsang, B., and Rao, S.M. (1997). Relationship between confabulation and measures of memory and executive function. *Journal of Clinical and Experimental Neuropsychology*, 19, 867–877.

Cusimano, C. (2012). *La Sémantique contemporaine du sème au thème*. Sorbonne Université Presses, Paris.

Cusimano, C. (2015). *Le Sens en mouvement – études de sémantique interprétative*, volume 11. Peter Lang, Frankfurt.

Cusimano, C. (2016). Réflexions critiques sur les tests lexicaux proposés aux malades d'Alzheimer. *Texto ! Textes & Cultures*, XXI(1) [Online]. Available: http://www.revue-texto.net/index.php?id=3762.

Dalla Barba, G. (1993a). Confabulation: Knowledge and recollective experience. *Cognitive Neuropsychology*, 10, 1–20.

Dalla Barba, G. (1993b). Different patterns of confabulation. *Cortex*, 29, 567–581.

Dalla Barba, G., Parlato, V., Lavarone, A., and Boller, B. (1995). Anosognosia, intrusions and "frontal" functions in Alzheimer's disease and depression. *Neuropsychologia*, 33, 247–259.

Dalla Barba, G., Mantovan, M.C., Cappelletti, J.Y., and Denes, G. (1998). Temporal gradient in confabulation. *Cortex*, 34, 417–426.

Dalla Barba, G., Nedjam, Z., and Dubois, B. (1999). Confabulation, executive functions, and source memory in Alzheimer's disease. *Cognitive Neuropsychology*, 16, 385–398.

Dartigues, J.-F., Frabrigoule, C., Lettenneur, L., Amieva, H., Thiessard, F., and Orgogozo, J.-M. (1997). Épidémiologie de la plainte mnésique. In *De la plainte mnésique à la maladie d'Alzheimer*, Michel, B.F., Dérouesné, C., and Gély-Nargeot, M.C. (eds). Solal, Marseille, 41–46.

Degiovani, R. (2014). Repenser la prise en charge orthophonique dans la maladie d'Alzheimer. *L'Orthophoniste*, 242, October.

Delacourte, A., David, J.-P., Ghozali, F., Fallet-Blanco, C., Vermersch, P., Wattez, A., Sergeant, N., Permanne, B., Buée, L., and Di Menza C. (1997). Lésions cérébrales de la maladie d'Alzheimer. In *De la plainte mnésique à la maladie d'Alzheimer*, Michel, B.F., Dérouesné, C., and Gély-Nargeot, M.C. (eds). Solal, Marseille, 29–46.

Derouesné, C. (1997). Méconnaissance des déficits mnésiques. In *De la plainte mnésique à la maladie d'Alzheimer*, Michel, B.F., Dérouesné, C., and Gély-Nargeot, M.C. (eds). Solal, Marseille, 57–66.

Derouesné, C. and Selmès, J. (2009). *La maladie d'Alzheimer pour les nuls*. First, Paris.

Devevey, A. and Balavoine, R. (2010). Analyse des productions implicites (néoformes) des patients au stade sévère des maladies neuro-dégénératives. In *Le langage oral : données actuelles et perspectives en orthophonie*, Rousseau, T. and Valette-Fruhinsholz, F. (eds). Ortho, Isbergues, 333–348.

Devevey, A. and Kunz, L. (eds) (2013). *Les troubles spécifiques de langage : pathologies ou variations ?* De Boeck/Solal, Brussels/Marseille.

Dierckx, E. (2008). *Early Detection of Alzheimer's Disease: A Neuropsychological Approach*. Academic & Scientific Publishers, Brussels.

Dubois, V. (2015). *Mémoire, ma petite mémoire – Alzheimer vu de l'intérieur*. Ker, Hévillers.

Dubois, B., Tounsi, H., and Deweer, B. (1997). Mémoire épisodique et sensibilité aux indices de rappel. In *De la plainte mnésique à la maladie d'Alzheimer*, Michel, B.F., Dérouesné, C., and Gély-Nargeot, M.C. (eds). Solal, Marseille, 153–157.

Emery, O.B. (1988). Language and memory processing in senile dementia Alzheimer's type. In *Language, Memory and Aging*, Light, L.H. and Burke, D.M. (eds). Cambridge University Press, New York, 221–243.

Eustache, F., Faure, S., and Desgranges, B. (2013). *Manuel de neuropsychologie*, 4th edition. Dunod, Paris.

Eynard, C. (ed.) (2012). *Alzheimer system : entre surmédiatisation de la maladie et invisibilité des personnes : changer notre regard...* Chronique sociale, Lyon.

Gagnepain, J. (2016). *Huit leçons d'introduction à la théorie de la médiation*. Institut Jean Gagnepain, Matecoulon-Montpeyroux [Online]. Available: http://www.institut-jean-gagnepain.fr/téléchargement/.

Galimberti, D. and Elio Scarpani, E. (eds) (2008). *Biomarkers for Early Diagnosis of Alzheimer's Disease*. Nova Biomedical Books, New York.

Gayraud, F. and Barkat-Defradas, M. (2012). Effet du voisinage phonologique sur l'accès lexical dans le discours spontané de patients Alzheimer. In *Actes de la conférence conjointe JEP-TALN-RECITAL 2012*, volume 1. JEP, 369–376 [Online]. Available: http://www.aclweb.org/anthology/F12-1047.

Guertin, M. (2000). *La conscience chez les personnes Alzheimer : étude de cas et éthique des comportements*. Médiaspaul, Quebec.

Hatfield, F.M. (1972). Looking for help from linguistics. *British Journal of Disorders and Communication*, 7, 64–81.

Hirsch, E. and Ollivet, C. (eds) (2007). *Repenser ensemble la maladie d'Alzheimer : éthique, soin et société*. Vuibert, Paris.

Hugonot-Diener, L., Barbeau, E., Michel, B.-F., Thomas-Antérion, C., and Robert, P. (2013). *Gremoire 1 : tests et échelles de la maladie d'Alzheimer et des syndromes apparentés*. De Boeck/Solal, Brussels/Marseille.

Hugonot-Diener, L., Thomas-Antérion, C., and Sellal, F. (2015). *Gremoire 2 : tests et échelles des maladies neurologiques avec symptomatologie cognitive*. De Boeck/Solal, Brussels/Marseille.

Isnard, G. (2009). *Survoltez votre mémoire – boostez vos neurones et votre concentration*. Ixelles, Brussels.

Jackson, C. (1996). *Understanding Psychological Testing*. Blackwell/The British Psychological Society, Oxford.

Jakobson, R. (1969). *Langage enfantin et aphasie*. Éditions de Minuit, Paris.

Jefferson, G. (2004). Glossary of transcript symbols with an introduction. In *Conversation Analysis. Studies from the First Generation*, Lerner, G. (ed.). Benjamins, Amsterdam, 13–31.

Kaplan, R.M. and Saccuzzo, D.P. (2005). *Psychological Testing – Principles, Applications, and Issues*, 6th edition. Thomson Wadsworth, Belmont.

Kline, P. (1993). *The Handbook of Psychological Testing*. Routledge, Abingdon-on-Thames.

Laurent, B. and Thomas-Antérion, C. (1997). Neuropsychologie de la maladie d'Alzheimer débutante. In *De la plainte mnésique à la maladie d'Alzheimer*, Michel, B.F., Dérouesné, C., and Gély-Nargeot, M.C. (eds). Solal, Marseille, 101–112.

Lee, H. (2012). Langage et maladie d'Alzheimer : analyse multidimensionnelle d'un discours pathologique. PhD Thesis, Université Paul-Valéry-Montpellier 3, Montpellier.

Leibing, A. and Cohen, L. (eds) (2006). *Thinking about Dementia: Culture, Loss, and the Anthropology of Senility*. Rutgers University Press, New Brunswick.

Leyhe, T., Saur, R., Eschweiler, G.W., and Milian, M. (2011). Impairment in Proverb interpretation as an executive function deficit in patients with amnestic mild cognitive impairment and early Alzheimer's disease. *Dementia and Geriatric Cognitive Disorders Extra*, 1(1), 51–61 [Online]. Available: https://www.ncbi.nlm.nih.gov/pmc/articles/PMC3199884/.

Lišková, E. (2017). La théorie des prototypes et tests lexicaux sur la mémoire : une analyse critique. Master's Thesis, Université Masaryk, Brno.

Mahieux, F. and Gély-Nargeot, M.C. (1997). Classification des troubles cognitifs du sujet âgé. In *De la plainte mnésique à la maladie d'Alzheimer*, Michel, B.F., Dérouesné, C., and Gély-Nargeot, M.C. (eds). Solal, Marseille, 85–98.

Malaguarnera, S. (2016). *Dictionnaire de neuropsychanalyse*. CreateSpace, Scotts Valley.

Marsaudon, E. (2008). *200 questions-clés sur la maladie d'Alzheimer – informer, traiter, accompagner, prendre en charge*. Ellébore, Paris.

Martin, P. (1993). *De la base cérébrale du fonctionnement phonique d'une langue*. Publications de l'Université de Laval, Quebec.

Maurer, K. and Maurer, U. (1999). *Alzheimer – vie d'un médecin. Histoire d'une maladie*. Michalon, Paris.

McClelland, J.L. (2000). Chapter 36: Connectionist models of memory. In *The Oxford Handbook of Memory*, Tulving, E. and Craik, F.I.M. (eds). Oxford University Press, New York.

Mjachký, J. (2016). Valeur et bi-axialité saussuriennes, théorie de la médiation, détermination nominale et nom propre : la structure (encore) en question. PhD Thesis, Université Masaryk, Brno.

Nori, G. and Warner, J. (2011). *Alzheimer et autres démences*. Modus Vivendi, Quebec.

Pancrazi-Boyer, M.-P. (1997). Mémoire épisodique et sensibilité aux indices de rappel. In *De la plainte mnésique à la maladie d'Alzheimer*, Michel, B.F., Dérouesné, C., and Gély-Nargeot, M.C. (eds). Solal, Marseille, 47–54.

Petersen, R.C. (2003). *Mild Cognitive Impairment: Aging to Alzheimer's Disease*. Oxford University Press, Oxford.

Piatek, D. and Desbons, M. (2009). *La Mémoire envolée*. Gecko, Paris.

Ploton, L. (2004). *Maladie d'Alzheimer – à l'écoute d'un langage*, 3rd edition. Chronique Sociale, Lyon.

Poujois, A., Perret-Liaudet, A., and Thomas-Antérion, C. (2010). Le piège de la maladie d'Alzheimer à présentation dysexécutive. *Revue de neuropsychologie*, 2(5), 28–31.

Quentel, J.-C. and Beaud, L. (2013). La déconstruction du langage à travers la théorie de la médiation. *Al-Lisaniyyat*, 25–39 [Online]. Available: https://hal shs.archives-ouvertes.fr/halshs-00909090.

Rastier, F. (1991). *Sémantique et recherches cognitives*. PUF, Paris.

Rastier, F. (2011). *La Mesure et le Grain. Sémantique de corpus*. Champion, Paris.

Rastier, F. (2012). Néologismes et néonazisme. Sur le diagnostic d'Anders Breivik. *Cités*, 50, 13–17.

Rosch, E. (1977). Human categorization. In *Advances in Cross-Cultural Psychology*, volume 1, Warren, N. (ed.). Academic Press, London, 1–49.

Rossi, J.-P. (2013). *Psycho-neurologie du langage – le sens du langage et des objets du monde*. De Boeck Université, Brussels.

de Rotrou, J. and Frambourt, A. (1997). Stimulation cognitive et plainte mnésique. In *De la plainte mnésique à la maladie d'Alzheimer*, Michel, B.F., Dérouesné, C., and Gély-Nargeot, M.C. (eds). Solal, Marseille, 197–209.

Rousseau, T. (2011). *Maladie d'Alzheimer et troubles de la communication. Évaluation, prise en charge thérapeutique*. Elsevier Masson, Issy-les-Moulineaux.

Rousseau, T. (2013). *Communiquer avec un proche Alzheimer. Comprendre, déculpabiliser et maintenir un lien*. Eyrolles, Paris.

Rowe, J.W. and Kahn, R.L. (1987). Human aging: Usual and successful. *Science*, 237(4811), 143–149.

Rumelhart, D.E. and Todd, P.M. (1993). Learning and connectionist representations. In *Attention and Performance XIV: Synergies in Experimental Psychology, Artificial Intelligence, and Cognitive Neuroscience*, Meyer, D.E. and Kornblum, S. (eds). MIT Press, Cambridge, 3–30.

Sabouraud, O., Gagnepain, J., and Sabouraud, A. (1963). Vers une approche linguistique des problèmes de l'aphasie. *Revue de neuro-psychiatrie de l'Ouest*, 87.

Sapir, E. (1970). *Le Langage – introduction à l'étude de la parole*. Payot, Paris.

Schnider, A. (2008). *The Confabulating Mind: How the Brain Creates Reality*. Oxford University Press, New York.

Spinelli, E. and Ferrand, L. (2005). *Psychologie du langage : l'écrit et le parlé, du signal à la signification*. Armand Colin, Paris.

Terry, R., Katzman, D., Bick, R.K., and Sisodia, S.S. (eds) (1999). *Alzheimer Disease*, 2nd edition. Lippincott Williams & Wilkins, Crawfordsville.

Thorndike, J. (2009). *Last of His Mind: A Year in the Shadow of Alzheimer's*. Ohio University Press, Athens.

Touchon, J. and Portet, F. (2004). *La maladie d'Alzheimer*. Elsevier Masson, Issy-les-Moulineaux.

Tran, T.M., Dassé, P., Letellier, L., Ljubinkovic, C., Théry, J., and Mackowiak, M.-A. (2012). Les troubles du langage inauguraux et démence : étude des troubles lexicaux auprès de 28 patients au stade débutant de la maladie d'Alzheimer. *Actes du Congrès mondial de linguistique française*, 1659–1672 [Online]. Available: http://dx.doi.org/10.1051/shsconf/20120100211.

Whorf, B.L. (1940). Science and linguistics. *Technology Review*, 6, 247–248.

Index

P, R, S, T

Other titles from

iSTE

in

Cognitive Science and Knowledge Management

2020

GILLIOZ Christelle, ZUFFEREY Sandrine
Introduction to Experimental Linguistics

NGUYEN-XUAN Anh
Cognitive Mechanisms of Learning (Learning, Development and Cognitive Technologies Set)

OSIURAK François
The Tool Instinct

ZUFFEREY Sandrine
Introduction to Corpus Linguistics

2019

CLAVEL Chloé
Opinion Analysis in Interactions: From Data Mining to Human-Agent Interaction

KAROUI Jihen, BENAMARA Farah, MORICEAU Véronique
Automatic Detection of Irony: Opinion Mining in Microblogs and Social Media

MARTINOT Claire, BOŠNJAK BOTICA Tomislava, GEROLIMICH Sonia, PAPROCKA-PIOTROWSKA Urszula
Reformulation and Acquisition of Linguistic Complexity: Crosslinguistic Perspective
(Interaction of Syntax and Semantics in Discourse Set – Volume 2)

2018

BONFANTE Guillaume, GUILLAUME Bruno, PERRIER Guy
Application of Graph Rewriting to Natural Language Processing
(Logic, Linguistics and Computer Science Set – Volume 1)

PENNEC Blandine
Discourse Readjustment(s) in Contemporary English
(Interaction of Syntax and Semantics in Discourse Set – Volume 1)

2017

KURDI Mohamed Zakaria
Natural Language Processing and Computational Linguistics 2: Semantics, Discourse and Applications

MAESSCHALCK Marc
Reflexive Governance for Research and Innovative Knowledge
(Responsible Research and Innovation Set – Volume 6)

PELLÉ Sophie
Business, Innovation and Responsibility
(Responsible Research and Innovation Set - Volume 7)

2016

BOUVARD Patricia, SUZANNE Hervé
Collective Intelligence Development in Business

CLERC Maureen, BOUGRAIN Laurent, LOTTE Fabien
Brain–Computer Interfaces 1: Foundations and Methods
Brain–Computer Interfaces 2: Technology and Applications

FORT Karën
Collaborative Annotation for Reliable Natural Language Processing

GIANNI Robert
Responsibility and Freedom
(Responsible Research and Innovation Set – Volume 2)

GRUNWALD Armin
The Hermeneutic Side of Responsible Research and Innovation
(Responsible Research and Innovation Set – Volume 5)

KURDI Mohamed Zakaria
Natural Language Processing and Computational Linguistics 1: Speech, Morphology and Syntax

LENOIR Virgil Cristian
Ethical Efficiency: Responsibility and Contingency
(Responsible Research and Innovation Set – Volume 1)

MATTA Nada, ATIFI Hassan, DUCELLIER Guillaume
Daily Knowledge Valuation in Organizations

NOUVEL Damien, EHRMANN Maud, ROSSET Sophie
Named Entities for Computational Linguistics

PELLÉ Sophie, REBER Bernard
From Ethical Review to Responsible Research and Innovation
(Responsible Research and Innovation Set - Volume 3)

REBER Bernard
Precautionary Principle, Pluralism and Deliberation
(Responsible Research and Innovation Set – Volume 4)

SILBERZTEIN Max
Formalizing Natural Languages: The NooJ Approach

2015

LAFOURCADE Mathieu, JOUBERT Alain, LE BRUN Nathalie
Games with a Purpose (GWAPs)

SAAD Inès, ROSENTHAL-SABROUX Camille, GARGOURI Faïez
Information Systems for Knowledge Management

2014

DELPECH Estelle Maryline
Comparable Corpora and Computer-assisted Translation

FARINAS DEL CERRO Luis, INOUE Katsumi
Logical Modeling of Biological Systems

MACHADO Carolina, DAVIM J. Paulo
Transfer and Management of Knowledge

TORRES-MORENO Juan-Manuel
Automatic Text Summarization

2013

TURENNE Nicolas
Knowledge Needs and Information Extraction: Towards an Artificial Consciousness

ZARATÉ Pascale
Tools for Collaborative Decision-Making

2011

DAVID Amos
Competitive Intelligence and Decision Problems

LÉVY Pierre
The Semantic Sphere: Computation, Cognition and Information Economy

LIGOZAT Gérard
Qualitative Spatial and Temporal Reasoning

PELACHAUD Catherine
Emotion-oriented Systems

QUONIAM Luc
Competitive Intelligence 2.0: Organization, Innovation and Territory

2010

ALBALATE Amparo, MINKER Wolfgang
Semi-Supervised and Unsupervised Machine Learning: Novel Strategies

BROSSAUD Claire, REBER Bernard
Digital Cognitive Technologies

2009

BOUYSSOU Denis, DUBOIS Didier, PIRLOT Marc, PRADE Henri
Decision-making Process

MARCHAL Alain
From Speech Physiology to Linguistic Phonetics

PRALET Cédric, SCHIEX Thomas, VERFAILLIE Gérard
Sequential Decision-Making Problems / Representation and Solution

SZÜCS Andras, TAIT Alan, VIDAL Martine, BERNATH Ulrich
Distance and E-learning in Transition

2008

MARIANI Joseph
Spoken Language Processing

www.ingramcontent.com/pod-product-compliance
Lightning Source LLC
LaVergne TN
LVHW011803200926
848484LV00003B/776
* 9 7 8 1 7 8 6 3 0 6 6 2 3 *